Handbook of Occupational Safety and Health

MW01534046

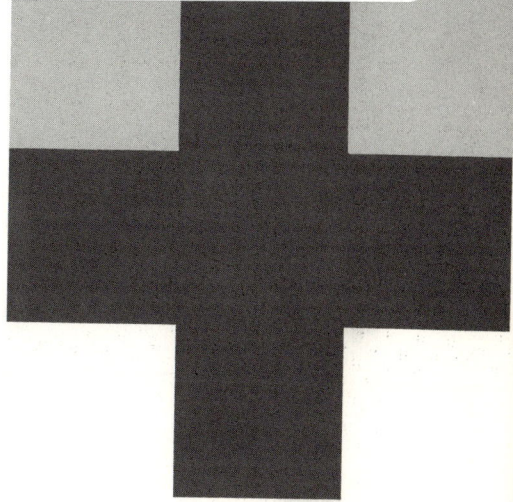

GREEN CROSS FOR SAFETY

NATIONAL SAFETY COUNCIL
Chicago, Illinois 60611

Third printing, 1979

International Standard Book Number: O-87912-036-3
Library of Congress Catalog Card Number: 75-11314
Printed in the United States of America

10M279   Stock No. 129.03

Foreword

This handbook is meant for owners, managers, and supervisors of smaller companies and small plants of large companies. Although the text is designed to be self-teaching, it can also be used in a basic safety course.

The handbook emphasizes setting up and running a safety, industrial hygiene, and health program.

• The first eight chapters tell what needs to be done and how to rank priorities in setting up a safety, industrial hygiene, and health program. Chapters discuss why safety is good business and how to get and maintain safe and healthful working conditions. Inspection techniques and how to follow up accidents to make sure they do not happen again are described. Accident records and reporting and OSHA considerations are discussed. Employee training and motivation are emphasized as being vital to an effective safety program.

• The second portion of the handbook stresses specifics—machine and tool safety, materials handling and storage, selection and use of personal protective equipment, fire protection, and basics of industrial hygiene.

Because this handbook is meant to be a compact introduction to the person with little safety and health experience, it does not go into great detail in any one subject. For the reader who has a specialized field of interest or a special problem, the National Safety Council has available a number of specialized books:

> *Accident Prevention Manual for Industrial Operations*
> *Communications for the Safety Professional*
> *Fundamentals of Industrial Hygiene*
> *Industrial Noise and Hearing Conservation*
> *Motor Fleet Safety Manual*
> *OSHA Standards Handbook for Small Business*
> *Supervisors Guide to Human Relations*
> *Supervisors Safety Manual*

In addition, NSC publishes approximately 400 Industrial Data Sheets on specialized subjects. (An alphabetical listing is available.)

In this handbook, references are made to "Title 29—Labor" of the

Code of Federal Regulations (29 CFR). These are the regulations promulgated by the U.S. Department of Labor under authority of the Occupational Safety and Health Act of 1970. They are first published in the *Federal Register* and are available in booklet form from the U.S. Government Printing Office, Washington, D.C., or from the area offices of the Occupational Safety and Health Administration.

Because of numerous references to these regulations, it is suggested that the reader obtain either the Occupational Safety and Health Standards for industry (29 CFR Part 1910) or the Safety and Health Regulations for Construction (29 CFR Part 1926), depending on whether industrial or construction work is being performed.

Although the text, in general, is a digest of the NSC publications listed earlier, much of the material was prepared by Occupational Safety and Health Consulting Associates, Inc., Silver Spring, Md.

The information and recommendations contained in this handbook have been compiled from sources believed to be reliable and to represent the best current opinion on the subject. No warranty, guarantee, or representation is made by the National Safety Council as to the absolute correctness or sufficiency of any representation contained in this and other publications, and the National Safety Council assumes no responsibility in connection therewith; nor can it be assumed that all acceptable safety measures are contained in this (and other) publications, or that other or additional measures may not be required under particular or exceptional conditions or circumstances.

This handbook will be revised periodically. Contributions and comments from readers are invited.

Contents

v

NATIONAL SAFETY COUNCIL

FOUNDED SEPTEMBER 24,1913
INCORPORATED IN ILLINOIS OCTOBER 1,1930
INCORPORATED BY ACT OF CONGRESS AUGUST 13,1953

Purposes and Powers. Certain essential provisions of the Act of Congress which incorporated the National Safety Council are as follows:

Objects and Purposes

"The objects and purposes of the corporation shall be--

to further, encourage; and promote methods and procedures leading to increased safety, protection, and health among employees and employers and among children, in industries, on farms, in schools and colleges, in homes, on streets and highways, in recreation, and in other public and private places;

to collect, correlate, .. and disseminate educational and informative data, .. relative to safety methods and procedures;

to arouse and maintain the interest of the people of the United States, its Territories and possessions in safety and in accident prevention, and to encourage the adoption and institution of safety methods by all persons, .. and .. organizations;

to organize, establish, and conduct programs, .. for the education of all persons, .. in safety methods and procedures;

to cooperate with, enlist, and develop the cooperation of and between all persons, .. and .. organizations .. both public and private, engaged or interested in, .. any or all of the foregoing purposes .."

Powers

"The corporation shall have power --

to establish and maintain offices for the conduct of its business, and to charter local, State, and regional safety organizations, .. in appropriate places throughout the United States, its Territories and possessions;

to charge and collect membership dues, subscription fees, and receive contributions or grants of money or property to be devoted to the carrying out of its purposes;

to choose such officers, directors, trustees, managers, agents, and employees as the business of the corporation may require;

to adopt, amend, and alter a constitution and bylaws, ..

to organize, establish, and conduct conferences on safety and accident prevention;

to publish magazines and other publications and materials, .. consistent with its corporate purposes;

to adopt, alter, use, and display such emblems, seals, and badges as it may adopt."

Nonpolitical Nature

"The corporation, and its officers, directors, and duly appointed agents as such, shall not contribute to or otherwise support or assist any political party or candidate for office."

No Stock or Dividends

"The corporation shall have no power to issue any shares of stock nor to declare nor pay any dividends."

Audit and Congressional Report

The financial transactions shall be audited annually, .. by an independent certified public accountant .. A report of such audit shall be made by the corporation to the Congress not later than six months following the close of such fiscal year for which the audit is made."

Exclusive Right to Name and Emblem

"The corporation, and its subordinate divisions and regional, state, and local chapters, shall have the sole and exclusive right to use the name, National Safety Council. The corporation shall have the exclusive and sole right to use, or to allow or refuse the use of, such emblems, seals, and badges as it may legally adopt .."

Transfer of Assets

"The corporation may acquire the assets of the National Safety Council, Incorporated, a corporation organized under the laws of the State of Illinois, upon discharging or satisfactorily providing for the payment and discharge of all of the liability of such corporation and upon complying with all laws of the State of Illinois applicable thereto."

Congressional and Presidential Approval. The Act which incorporated the National Safety Council was passed during the First Session of the 83rd Congress and was designated Public Law 259. Leadership in the Congress was as follows:

In the Senate	Bill introduced by Arthur V. Watkins Senator from Utah	Approved by the Sub-Committee on Federal Charters of the Judiciary Committee, Sub-Committee Chairman, John Marshall Butler Senator from Maryland	Approved by the Judiciary Committee, Chairman, William Langer Senator from North Dakota	Passed by the Senate, President of the Senate, Richard M. Nixon Vice President of the United States
In the House of Representatives	Bill introduced by Clifford Davis Representative from Tennessee	Approved by the Sub-Committee on Federal Charters of the Judiciary Committee, Acting Sub-Committee Chairman, John Robsion Representative from Kentucky	Approved by the Judiciary Committee, Chairman, Chauncey W. Reed Representative from Illinois	Passed by the House, Speaker of the House, Joseph W. Martin, Jr. Representative from Massachusetts

The Act was signed by the President of the United States, Dwight D. Eisenhower, on August 13, 1953.

Transfer of Operations to the Federal Corporation. At the time the Act was passed granting the National Safety Council a federal charter, the Council was functioning as an Illinois corporation. This arrangement continued until January 1, 1954 on which date the Council's assets, operations and organizational structure were transferred from the Illinois corporation to the federal corporation.

The National Safety Council is the only national nonpolitical and nonprofit organization established solely for accident prevention. In 1953, the 83rd Congress of the United States issued a federal charter for the Council. This recognized it as an integral part of the American way of life.

Introduction

Cost-conscious companies find that they must control accident costs if they are to continue in a highly competitive market. Safety must not be considered a nuisance but a necessity—accident prevention is not a fringe benefit but a prerequisite to a profitable operation.

Activities related to safeguarding the company's investment and continuity of operation need to be coordinated with the safety effort. Since OSHA, the Consumer Product Safety Act, the Environmental Protection Acts, and the Federal Energy Program, this is all the more important. All safety and health considerations must be closely integrated.

Basic Elements of a Safety Program

Basic to any program is organization—the method employed by management to assign responsibility for accident prevention and to assure performance under that responsibility. This is outlined here and more fully explained in later chapters.

Analysis of successful programs in plants of all sizes and organizations in all industries shows that the programs are built around seven basic elements:

1. Declaration of management policy and leadership
2. Assignment of responsibility, authority, and accountability
3. Maintenance of safe working conditions
4. Establishment of safety training
5. Establishment of an accident reporting and analysis system
6. Creation of medical and first aid programs
7. Acceptance of personal accountability by employees

Management policy and leadership

Top management's attitude toward accident prevention is almost invariably reflected in the attitude of the supervisory force. Similarly, the worker's attitude is usually the same as his supervisor's. Thus, if the top executive is not genuinely interested in preventing accidents and injuries, no one else is likely to be. Since this basic truth applies to every level of management and supervision, an accident prevention program must have top management's announced and demonstrated interest if employee cooperation and participation are to be obtained.

Other reasons for having a declared safety policy are:

1

• A good policy makes it easier to enforce safe practices and maintain safe and healthful conditions.

• It makes it easier for supervisors to comply with company policy.

• It makes it easier for employees to follow safety instructions.

• It makes it easier to select and purchase the proper equipment and to obtain good preventive maintenance of equipment.

Basic to a safety policy are these statements:

• The safety of employees and the public is paramount.

• Safety will take precedence over expediency.

• Every attempt will be made to reduce the possibility of accidents.

• The company intends to comply with all safety laws and ordinances.

Once a safety policy is established, it should be posted and publicized. However, its effectiveness will vary directly with the active support given it by management.

Assignment of responsibility, authority, and accountability

All operating managers and supervisors must be assigned specific responsibilities for their part of a safe and healthful workplace; they must be given authority to carry out their responsibilities, and must be held accountable for the results. Supervisors of operating units must be measured by their safety record as well as by their production.

The supervisor is the key in any safety program. To the employee, he *is* management. The thoughts and plans of the top executive must reach the workers by way of their supervisors.

Active management and control of the safety program may be vested in the top executive, the general manager, or in an experienced and qualified supervisor who has both authority and status. Whoever is responsible must be familiar with state and federal safety and health standards and regulations.

Organizations with scattered operating units, such as construction, repair, service, and trade units, face special safety problems. Dispatch centers and other locations from which the various units operate should be used as bases for safety and health promotion.

If representatives from the different operating units of a business meet regularly to discuss the detection and control of unsafe conditions and acts, they can do much to keep the safety program productive.

Maintenance of safe working conditions

Before any control measures can be set up, the hazards in the work operation must be identified. Federal, state, and voluntary safety and health standards can be used as a measure of what should be the minimum requirements.

Identification of hazards requires a fairly complete inspection of all operating areas. The

Industrial hygienist with velometer checks air flow over a paint dip tank. His colleague *(left)* checks for flammable vapor.

inspection can be made by a small group, such as one consisting of someone from production, someone from maintenance, and the person responsible for coordinating the safety program. Insurance company engineers can also be helpful, particularly on an initial inspection. The inspection can be done in stages so as not to interfere with the normal routine.

Safety inspection checklists may be helpful in directing the inspection to key operating areas or to the most hazardous exposures. Checklists are available from insurance companies and the National Safety Council.

After an initial inspection has been made and all deficiencies listed, a program for the correction of those deficiencies and for regular and periodic inspections must be set up. Finding unsafe conditions by means of inspection and promptly correcting them is one of the best methods for management to demonstrate to employees its interest and sincerity in accident prevention.

Whenever possible, make corrections right away. Many inju-

3

ries are caused by seemingly minor things—slippery floors, defective hand tools, missing guards, and poor housekeeping.

Periodic inspections are essential:

• To make sure that new or recurring hazards are identified

• To make sure that equipment is being operated in a safe manner, that personal protective equipment is being used, and that processes and equipment are under control

• To check on general housekeeping, first aid facilities, the condition of fire fighting equipment, and the condition of storage areas

Certain hazards may not be recognized at first because there is no record that they have caused injuries or illness, or because their presence can be detected only by trained specialists. In order to assure maximum coverage, supervisors and insurance company safety engineers should be consulted and previous accident reports checked. All shifts of an operation should receive the same attention. More details are in Chapter 4, Safety Inspections.

Because certain solvents and chemicals may endanger the health of employees, create fire and explosion hazards, or react with other chemicals, the hazardous properties of all solvents and chemicals should be determined and identified before they are approved for use in production. If in doubt, get outside expert help. See Chapter 10, Materials Han-

dling and Storage, and Chapter 13, Industrial Hygiene.

Establishment of safety training

No matter how well safety is engineered into an operation, much of the safety of employees depends upon their own conduct. Some people work safely in dangerous surroundings, whereas others have accidents on jobs that seem quite safe. Controlling people is, therefore, a necessary part of an accident prevention program.

It is necessary to influence the voluntary acts of workers by education and motivation. Much of the success obtained from safety and health activities is determined by how well workers are trained and influenced.

The wise employer takes advantage of this situation and through job safety analysis and training sees to it that each employee learns correctly those things that will make him a safe and efficient worker. (Another benefit of job safety analysis is hazard identification.)

Safety training is not difficult if handled as part of ordinary production training. It is always easier to train employees to form safe habits while they are learning their jobs than it is to correct established unsafe habits later.

If the job training procedures outlined in Chapter 7, Training, are followed, employees will be trained for efficient—as well as safe—operation.

Training results in fewer mistakes, less spoilage, fewer injuries, fewer delays, and lower overall production costs. In addi-

4

tion, less supervision will be necessary in the long run.

Establishment of an accident reporting and analysis system

The average smaller organization may not have enough disabling injuries in a month to warrant an extensive accident reporting or analysis system. In a year there may be only a few compensation or lost-time cases. When compared with the total number of manhours worked, a relatively few cases can produce an alarmingly high accident rate. This means unnecessarily high insurance rates and may even mean the difference between a credit and debit insurance rating.

It is quite important that a company keep some simple records. A high incidence of minor eye injuries, for example, certainly indicates that either the source should be eliminated or protective glasses or faceshields worn. Failure to act on these warnings will eventually result in expensive medical treatment, lost time, and perhaps even permanent disabilities. Analyzing simple accident records will provide convenient and systematized warnings. See Chapter 5, Accident Records and Reports.

Creation of medical and first aid programs

Medical and first aid programs, regardless of size, are composed of elements and services designed to *(a)* maintain the health of the work force, *(b)* prevent or control occupational and nonoccupational diseases and accidents, and *(c)* to prevent and lower disability and the resulting lost time.

Preplacement medical examinations should be made to determine and record the physical condition of the prospective worker so that he can be assigned to a suitable job in accordance with his mental ability and physical capacity. Hearing disabilites may be hazardous in some jobs. Vision disabilities could create a hazard where depth and field of vision are important. Previous exposure to radiation, asbestos, and carcinogens may be hazardous in a job that could aggravate a condition.

Periodic examinations may be necessary for workers who are regularly exposed to processes or materials hazardous to health or whose work involves responsibility for the safety of others, such as vehicle operators. Lead, silica, asbestos, and other substances are capable of causing diseases if the exposure level is not controlled. The frequency of the examinations must vary in accordance with the quality of the engineering control, the nature of the exposure (the rapidity of the action of the hazardous substance on the human body), and the findings of each examination. If a chronic health condition is revealed it may be necessary to restrict an employee to certain work.

Some organizations give examinations upon termination of employment, particularly where operations involve definite exposures to harmful noise levels or to health-hazardous substances, such as lead, benzene (benzol), and silica and asbestos dusts.

Urine and blood tests, chest X-rays, and vision and hearing tests may be a good investment and

5

eliminate controversies over whether a health condition is job related and to what extent. Health standards will be expanded and more rigorously enforced by OSHA as it expands and matures. The pressures developed by consumer groups, labor organizations, environmentalists, and others are forcing government to exert an ever-increasing control over safety and health risks from any source.

Because of the severe penalties which may be levied, no employer can afford to ignore the legislative mandates. In order to assure that operations do not result in violations, an organized approach to the control of safety and health hazards must be established and be given top priority.

In most small organizations and in field operations, it is not practical to have full-time professional medical personnel available. In such cases, the best arrangement is to have a suitable first aid kit or station administered by trained first aiders who follow procedures and treatments outlined by a doctor or who have Red Cross first aid training.

Regardless of the type, size, and extent of the operation, every employer must have posted, a step-by-step medical and first aid procedure to be followed in case of emergency. Emergency transportation to a hospital must also be available to transport injured or ill parties.

Acceptance of personal accountability by employees

The passing of laws, the promulgation of safety and health regulations, and the threat of penalties for violations will not of themselves assure safety at the workplace. Much of the safety of employees depends upon their own conduct and upon management interest and example.

An employee's conduct is framed by the type of training in safe practices he has received and the types of attitudes he has formed. Many executives feel that safety training is the best means of influencing attitudes and bolster safety training with contests, suggestion systems, awards, posters, slogans, and other means for maintaining employee interest. See Chapter 8, Promotion, Motivation, and Employee Involvement.

Employees will not accept personal accountability unless management takes the lead by:

• Setting a good example. For instance, if operations require employees to wear goggles or other personal protective equipment in certain areas, bosses should conform to these regulations when they visit these areas.

• Regular attendance at safety meetings.

• Reviewing and acting upon the safety records (good or bad) of different operating units.

• Making sure that every production discussion recognizes that safe operation is efficient operation.

• Demonstrating an interest through notices, bulletin board announcements, and other publicity.

6

Safety and Good Business

CHAPTER 1

Why Have a Safety and Health Program?

The federal Occupational Safety and Health Act of 1970.

Every employer whose business "affects commerce" is subject to the federal Occupational Safety and Health Act (OSHAct) of 1970. A business "affects commerce" if any of the tools, equipment, materials, or devices used in it were manufactured in another state. This means that virtually every employer in every industry is covered, which includes some 5 million employers and 60 million workers.

The OSHAct establishes two responsibilities for employers. First, they must provide employment and places of employment free from recognized hazards (a general duty responsibility); and second, they must comply with safety and health standards published by the U.S. Department of Labor.

Employees also have responsibilities under the OSHAct. Employees must comply with all of the rules, regulations, and standards applicable to their actions and conduct.

Employers can be penalized through government civil or criminal action if they violate their responsibilities. Employees cannot be penalized by OSHA (the Occupational Safety and Health Administration) for violating their responsibilities. Therefore, this means that meeting the OSHA safety and health standards, rules, and regulations must be made a condition of employment for every worker in every union contract or in every nonunion working agreement. Not only should meeting these safety and health requirements be made a condition of employment, but action must be taken by employers to make sure that these requirements are enforced by supervision at the worksite during operations.

The OSHAct provides that the federal government may make agreements with state governments to inspect work operations and enforce safety and health requirements if the state develops a program which is as effective as the federal program. To be "as effective as the federal program" does not mean the state must have identical standards, rules, and

7

regulations and identical enforcement organizations and personnel. It does mean, however, that if a state agrees to accept the responsibility through a formal legal agreement, it cannot treat employers more lightly than the federal government might. The state program is subject to scrutiny and evaluation, and if it does not measure up to be "as effective as" the federal program, the agreement can be terminated.

So, whether the federal or state inspector has jurisdiction over workplaces in a particular state, employers must have a safety and health program to make sure that the published standards are met and the place of employment is as free from recognized hazards as possible.

The Consumer Product Safety Act of 1972

In addition to promulgating a law to protect the safety and health of workers, Congress has passed legislation to protect the safety and health of consumers.

A National Commission on Product Safety was appointed in 1968 to conduct a comprehensive study investigating the scope and adequacy of measures being employed to protect consumers against unreasonable risks of injury that might be caused by hazardous consumer products. In its final report to Congress, the Commission estimated that 20 million Americans were injured each year as a result of incidents connected with household consumer products—30,000 were killed and 110,000 permanently disabled. The annual cost to the nation

was more than $5.5 billion. It was estimated that 20 percent of the injuries could have been prevented.

The Consumer Product Safety Act (CPS Act) was passed by Congress on October 12, 1972, and became effective on December 26, 1972. As a result, consumer products became subject to federal regulations and the government was given the authority to set safety and health standards for products and to force recall or correction of, or to actually ban, those products that present a real, substantial hazard to consumers.

The CPS Act joins a growing list of worker, consumer, and citizen protection laws passed during the past ten years which reflect the nation's concern with its safety and health environment. Each such act puts an additional burden on the employer to make sure that his business activities do not violate federal regulations.

A consumer product, subject to the CPS Act, is any article or component part of an article, produced or distributed for sale for personal use, consumption, or enjoyment in or around a household or residence, a school, or recreation area. Such products include all retail products used by consumers in or around a household (except foods, drugs, cosmetics, motor vehicles, tobacco products, pesticides, firearms and ammunition, radiological equipment and products, boats and aircraft, and certain flammable fabrics; these are covered by other laws).

Safety and health standards developed under the OSHAct may overlap those developed under the CPS Act. Employers who produce a product that is subject to the CPS Act must be aware of this fact.

Manufacturers, service organizations, distributors, construction firms, or private labelers, must establish a specific program to guide and protect their actions. Power lawnmower, ladder, and portable tool manufacturers are three industries that have already done so. It is suggested that a safety and health program be devised which encompasses both employee and product (consumer) safety and health factors.

To minimize difficulties in meeting both areas of responsibility, and to help establish the most effective management tools for efficiency and profitability, the following elements, each with hazard potential and control built into them, should be included in any program.

1. Operating policies should be in writing and should establish management guidelines for all operations.

2. Job and process standards and procedures should be in writing, as specific as possible, and included in employee training and supervisory responsibilities.

3. Design specifications, certifications, and quality control evaluation standards must be established and rigorously followed.

4. Purchasing specifications must be established with consideration for worker and consumer product hazard potential.

5. Receiving, handling, storage, and shipping procedures must assure that materials and products meet safety and health specifications in and out of the workplace.

6. Labels, warnings, instructions, and operating control standards must be applied wherever foreseeable hazards may exist.

7. Maintenance, repair, and installation specifications must be established.

8. Records of operations, quality control, receiving, storage, and shipping must be established and maintained to provide a means of review and followup in case problems occur.

9. An internal communications and feedback system should assure proper movement of information down the line and good or bad performance information back up the line.

10. Speedy investigative and complaint handling procedures should be established.

11. Advertising, merchandizing, and warranting claims must be monitored.

12. Contractual agreements and responsibilities must be clearly defined for all parties.

Having a management program incorporating these twelve operating elements of control will not only improve the profitability of the operation, but will also result in safety and health benefits for both employees and consumers.

Environmental requirements

A national policy of protection conservation, and enhancement of man's environment was established in 1969 by the National Environmental Policy Act (NEPA). It provides for a continuing policy of the federal government to use all practical means and measures to create and maintain conditions under which man and nature can exist in productive harmony.

The NEPA requires federal agencies to prepare detailed reports of the environmental consequences before authorizing any major actions which might significantly affect the quality of the human environment. This forces management to consider the environmental aspects of its proposed actions.

An employer, in protecting the safety and health of his workers on the job, may violate the environmental safety and health of the citizens living in the area if he is not aware of his responsibilities. For example:

1. Process air and air contaminants in a factory, construction site, or service establishment may be dangerous to workers and so is exhausted in accordance with safety and health standards. However, in the process of exhausting the contaminants from the working area, an exposure to those outside the working area may be created.

2. Dangerous or nuisance materials used or created in industrial processes, in service operations, or in construction or demolition activities may be flushed or washed out of the working areas into water or sewage systems, and thus, the hazard may be transferred from workers to those in the general environment.

3. Waste and scrap materials from work processes or work areas must be disposed of or reprocessed without creating undue hazards to workers or to others in the area.

4. Noisy work operations cannot expediently be moved so as to transfer the exposure from the worker to the citizenry.

The point of these examples is that the employer must be aware of the interrelationship of his mandated responsibilities and must keep in balance the sometimes-conflicting interests of the workers, consumers, and the environment.

If an employer or manager does not establish programs incorporating controls in all of these interrelated areas, he may run afoul of the law. Merely having a great safety and health program just for workers or just for consumer products or just for the environment is no longer acceptable.

Energy problems

As employers try to cope with the new energy problems and the pronouncements of the Federal Energy Office, safety and health factors must also be considered. Well-intentioned actions to conserve energy may indirectly create undue hazards to workers. For example:

1. Reduced illumination levels in

work, storage, exit, parking, and other areas may indirectly contribute to accidents and poor workmanship.

2. Reduced temperature levels may indirectly contribute to accidents, illnesses, and poor workmanship of certain materials.

3. Reduced voltage levels may cause certain instruments and equipment to operate improperly or operate erratically, and contribute to poor quality and accident potential.

4. Reduced ventilation to save power or heat for makeup air may cause harmful dust, vapor, and mist concentrations to build up beyond allowable levels. This may create health, fire, and explosion hazards.

5. Reduced power, heat, or energy application to a process may result in excessive levels of air, liquid, or solid pollution being released into the environment.

The point is that conservation actions should only be initiated after a thorough listing and evaluation of related (and even peripheral) effects.

An employer must have a program to cover the broad range of safety and health exposures which may be created by his business activities.

Conservation of Human and Material Resources

Costs to the nation

In 1973, out of 85,300,000 U.S. workers in all industries, 14,200 (0.016 percent) were killed and 2,500,000 (2.9 percent) suffered disabling injuries due to work accidents.

In 1974, out of 87 million workers, 13,400 (0.015 percent) were killed and 2,300,000 (2.6 percent) suffered work-related disabling injuries.

With an estimated ¼ billion lost man-days and $14 billion direct cost to the nation each year from workplace accidents, is it any wonder that Congress became concerned and passed the OSHAct to require employers to initiate safety and health efforts to control hazards on the job?

But the OSHAct was not the first effort to make a safer environment. The National Safety Council was established in 1913. Member companies of the National Safety Council have historically had better safety programs than nonmember companies and have had generally lower injury rates. See Table 1-A.

OSHA and the Bureau of Labor Statistics compile injury statistics on a different basis than does the NSC. The American National Standard *Method of Recording and Measuring Work Injury Experience,* Z16.1-1973 (R 1967), is used by those companies reporting to the National Safety Council. This is fully detailed in Chapter 5, Accident Records and Reports.

The Bureau of Labor Statistics, U.S. Department of Labor, collects injury statistics under the OSHAct. The first full calendar year of an employer's recordkeeping experience under OSHA covered 1972. The results for 1972 as to Recordable Occupational Injury and Illness Incidence Rates by Size of Unit and Industry Divi-

TABLE 1-A
COMPARATIVE INJURY RATES—NSC MEMBERS AND NONMEMBERS

Year	Injury Frequency Rate			Injury Severity Rate		
	NSC Members	Non-members	% NSC Rates Lower	NSC Members	Non-members	% NSC Rates Lower
			Manufacturing			
1965	4.6	16.8	−73	450	830	−46
1966	5.1	17.6	−71	470	810	−42
1967	5.1	18.0	−72	460	830	−45
1968	5.3	17.9	−70	480	790	−39
1969	5.7	18.9	−70	500	830	−40
1970	6.0	19.1	−69	500	860	−42
			Mining			
1965	23.5	37.9	−38	4,024	6,970	−42
1966	21.9	38.4	−43	4,016	6,534	−39
1967	22.4	37.5	−40	4,019	6,092	−34
1968	22.2	36.7	−40	4,494	7,983	−44
1969	21.8	37.6	−42	3,597	6,506	−45
1970	22.2	39.2	−43	3,793	6,644	−43

Source: Manufacturing nonmember—Total U.S. experience projected from BLS rates, less experience of reporters to NSC. Mining nonmember—Industrywide rates from Bureau of Mines, less reporters in U.S. Bureau of Mines safety competitions.

sion is shown in Table 1-B.

OSHA incidence rates are based on cases per 100 man-years of work (200,000 man-hours), while Z16.1 frequency rates are based on cases per 1,000,000 man-hours of work.

These figures indicate that although smaller business employers may not be able to afford full-time safety and health professionals, they have on-the-job work injury problems which need attention. When added to consumer product safety and health, environmental safety and health, and energy problems, a much stronger case for management concern and attention to safety and health responsibilities is made.

Costs to the employer

Many managers and employers are so accustomed to careless operating procedures concerning safety and health matters that they oppose regulatory standards which could make their operations much more profitable. They look only at the costs without considering the potential returns and the profit protection that a good accident prevention program provides.

Employees will develop slower work habits to compensate for hazardous situations facing them on the job. These can add up to significant hidden operating costs.

For example, if an employer

TABLE 1-B
RECORDABLE* OCCUPATIONAL INJURY AND ILLNESS
INCIDENCE RATES**
BY SIZE OF UNIT AND INDUSTRY DIVISION, 1972

Industry Division	*Number of Employees*								
	All Sizes	*1 to 19*	*20 to 49*	*50 to 99*	*100 to 249*	*250 to 499*	*500 to 999*	*1,000– 2,499*	*2,500 & over*
Private nonfarm sector†	10.9	5.7	10.3	13.3	14.7	13.7	12.2	10.9	11.1
Contract construction	19.0	14.3	19.8	22.8	24.9	24.2	19.7	15.1	12.7
Manufacturing	15.6	11.8	16.5	19.5	20.2	17.3	14.3	11.9	12.4
Transportation, public utilities	10.8	8.0	12.7	12.6	11.6	9.0	10.0	11.3	11.4
Wholesale and retail trade	8.4	4.6	8.9	11.1	12.1	11.5	12.4	11.9	10.2
Services	6.1	2.7	4.6	7.6	7.9	8.4	9.4	8.9	6.4
Finance, insurance, real estate	2.5	2.1	1.9	2.9	2.8	3.0	2.9	2.8	1.8

Source: Bureau of Labor Statistics, U.S. Department of Labor
Includes fatalities, lost workday cases, and nonfatal cases without lost workdays.

$$**Incidence\ Rate = \frac{No.\ of\ injuries\ \&\ illnesses \times 200,000\ (100\ full\text{-}time\ equivalent\ workers \times 40\ hrs/wk \times 50\ wks/yr)}{Total\ hours\ worked\ by\ all\ employees\ during\ period\ covered}$$

†*Excludes railroads and mining.*

pays an employee $6.00 per hour to do certain productive work, how much of this amount should be spent paying him to dodge hazards?

• How much production time is wasted by workers dodging holes, scrap piles, debris, blocked aisles, and spills in addition to doing their jobs?

• How much production time is wasted when a worker cannot breathe properly due to nuisance or dangerous dusts or vapors which are not properly exhausted from his work area?

• How much production time does a worker on a scaffold without a guardrail waste worrying about whether he might step back and fall off?

• Is it reasonable to expect a worker on an unguarded machine to use part of his production time worrying whether he might get caught in the rotating gears, the ram that is moving up and down, or the reciprocating arms? He might feed the machine less quickly, vigorously, and productively than if it were guarded and he did not have his attention diverted by such fears.

• How much production time is lost due to the distractions of noise, vibration, glare, improper illumination, and temperature extremes in a work area?

13

Wouldn't a worker be more productive if these accident producing conditions did not exist and he did not have to compensate for them in addition to doing his job?

If the $6.00 per hour worker spends *only two minutes per hour* worrying about his safety or dodging hazards just to get his work done in spite of them, *it costs the employer $400 per year* just to pay for this wasted time. In a 100-employee organization, this can total up to a hidden cost of $40,000 per year—lost just because safety and health conditions are not kept under control.

If these same conditions are not always completely compensated for by employees and result in injuries, the more visible costs to employers become apparent. Workers' compensation costs, medical costs, lost time in the production process by all employees exposed to the accident, possible damage to machines, materials, and equipment, and other costs may be more easily computed.

There are many obvious and direct accident costs to employers. But, the hidden costs, described by these examples, of operational inefficiencies which may cause accidents if not corrected may be even greater and are more insidious.

If a 100-employee organization can afford to spend $40,000 per year to dodge hazards, it may be more profitable to spend some or all of this amount to correct the hazards in the first place.

Costs to the worker

Workers always lose when they are involved in an accident which results in their injury. There is just no way they can be compensated completely for their losses. Although workers' compensation insurance takes care of the direct or obvious costs of injuries to workers (mainly medical costs and lost wages), serious and permanently disabling cases are another matter. In addition to the obvious costs, there are many other factors which can never be properly compensated for. These may include:

• Pain and suffering to the injured

• Worry and stress on the family and friends

• Loss of skills or abilities to continue a trade or a career

• Disruption of family, social, economic, and other routines

• Changes in life styles, goals, and objectives of the injured and his family

• A whole new life pattern for the family in fatal cases.

Since the worker is seldom able to control his work environment completely, he must either learn to compensate for the hazards he is exposed to, bring them to his employer's attention and hope they will be corrected, work through his labor union to get them corrected, or try to get the government to exert its pressure on his behalf.

As a result of many employers not recognizing the hazards in the workplace or not being willing to voluntarily correct them, pressures were exerted in Congress to pass the OSHAct. Now, the government must see that minimum safety and health standards are met in every workplace.

The government and employers alone will not prevent injuries to employees. Employees must continually cooperate, since every hazard cannot be feasibly controlled by legislation or by engineering means. Their actions and conduct must be such as to supplement and complement employer and governmental controls—or they will always lose in the end.

Control Injuries—Control Costs

Safety and health priorities as a control

The federal Occupational Safety and Health Act, the Consumer Product Safety Act, and the Environmental Policy Act will force employers to develop and utilize the most imaginative and advanced techniques of management and technology to give safety and health matters a higher priority in the scheme of things.

Before the OSHAct, there was no national safety and health code for employers to be measured against. Now there is, and it establishes a new national standard of conduct for safety and health on the job.

Before the CPS Act, there was no national safety and health code for manufacturers to be measured against. Now there is a national standard of conduct for manufacturers of products and for all others in the distribution chain.

Before the Environmental Policy Act, there was no national policy for the protection and improvement of man's environment. Now there is, and every business whose activities might affect the safety or health of man's environment will be subject to prosecution and penalty.

Business can no longer accept safety and health risks, on the job or elsewhere, which other generations considered "unavoidable." And, the acceptability of risk must be resolved in the context of rapidly changing value systems. The community image of a firm that is "a good neighbor and a safe place to work" is a major asset.

Since small businesses usually cannot justify a full-time safety and health professional, the whole concept of hazard control must be integrated with the normal management techniques for controlling operations. This is not difficult to do and can usually be accomplished by considering certain people factors, work area factors, work operation factors, and management control factors.

Controlling "people factors"

In producing a product, constructing a structure, or providing a service, how people are obtained, employed, and maintained is a prime concern of management. If certain basic management techniques are applied, jobs are done more efficiently and profitably—and safety is a byproduct. For example:

• If job standards and specifications are not developed, management will not know what type of person to hire for the job—and quality, quantity, and safety may be affected.

15

• If people are not selected and placed on the job with due consideration given to their physical and mental abilities as measured against particular job standards and specifications, they will not produce products of maximum quantity and quality; they may get injured or damage productive resources.

• If people are not educated and trained thoroughly in the job standards and procedures, they will make mistakes, damage the product, delay the work, and possibly hurt themselves.

• If people are not instructed or shown how their jobs are significant and fit into the overall scheme of things, they may not be as interested and motivated in producing to the best of their ability.

• When people are not supervised reasonably and knowledgeably by someone they can respect or relate to, they may develop or reinforce personal attitudes and hangups which may be counter productive from a production, safety, and health aspect.

• If people are not compensated, motivated, disciplined, and rewarded in a reasonable and practical manner, the job and safety controls may suffer.

By analyzing these six "people factors," it should be evident that using a more scientific approach to selecting, using, and maintaining workers may result in better management control and profitability. Safety and health and profit protection are spin-off benefits.

16

Controlling work area factors

In addition to scientifically dealing with people on the job, their work area must be considered from the operational and employee standpoint. For example:

• If machines and work areas are not arranged to accommodate the most efficient flow of materials to and from the work stations, there will be delays due to unnecessary handling, bottlenecks, confusion, and stress—as well as possible injuries due to bumping and hitting objects, tripping, and improper materials handling.

• If holes in the floor, scrap, spills, work in process, or other objects create obstacle course situations for a worker, then quality, quantity, and safety will be affected.

• If a worker must use one of his hands to hang onto something, or worry about contacting an energized circuit, or be concerned about falling to another level—in addition to doing his job—he cannot produce at maximum quantity and quality and eventually may be injured.

• If aisles, exits, and other means of access or egress are not marked, kept clear and protected, property may be destroyed and employees injured if an emergency situation arises.

• If proper illumination, ventilation, temperature, and humidity conditions are not maintained, the safety and health of employees may be affected.

This is not a complete listing of work area factors but is intended to indicate the basic concept involved. A different mixture of factors would be inherent in construction work, trade and service activities, and manufacturing.

Controlling work operation factors

In addition to people and work area factors, there are also work operation factors to be considered. For example:

• If materials are used, processed, stored, or handled so that fumes, dusts, vapors, heat, noise, vibrations, radiations, and other stresses foreign to the human body are created beyond acceptable levels, the quantity and quality of work will suffer and safety and health problems will result.

• If machines and rotating mechanisms are left unguarded, if defective electrical power tools are used, if no protection from flying or falling objects is provided, if radiation equipment is left unsafeguarded—eventually an employee will be injured or contract an occupational illness.

• If toxic, noxious, flammable, or explosive substances and conditions are free to be breathed, ignited, or contacted by employees, eventually they will be.

• If chemicals are free to contaminate, irritate, corrode, consume, or otherwise do damage, eventually they will do so.

• If machines and tools are not maintained properly or are used for operations other than that for which they were designed or are operated by persons not trained or qualified in operating them, eventually an employee will be injured.

The quality of management and supervision, which cuts across all operational, work area, and people factors, is the key or catalyst to pulling everything together in the most effective way.

Management control factors

The manager (or employer) is, of course, the brain of the whole control system. How he pulls together, organizes, and directs all the factors involved will determine the ultimate profitability and safety of the endeavor. Some managers are better at doing these things than others.

Let us review some common management weaknesses and their specific remedies.

1. Assuming people know company policy. *Remedy*—put it in writing.
2. Having generalized goals and objectives. *Remedy*—spell them out in detail.
3. Expecting people to know the limit of their authority and responsibility. *Remedy*—spell it out in the job description.
4. Assuming that if a worker has experience, he should be able to do the job. *Remedy*—establish set job standards and procedures and see that they are followed.
5. Making assignments and assuming that the results will take care of themselves. *Remedy*—establish accountability for results and regularly and periodically evaluate results.
6. "The design and procedure have been good for 20 years, they ought to be OK now."

17

Remedy—establish a means to be aware of new technology, new materials, and new techniques.

7. "Word of mouth direction is OK." *Remedy*—put things in writing.

8. "Someone will tell me if something goes wrong." *Remedy*—have a definite recording and reporting procedure.

9. "We made 10,000 last month so we should be able to do as well this month." *Remedy*—have a definite system of measuring and reporting results before the last minute.

10. Assuming conditions will never change. *Remedy*—reappraise and reevaluate contracts and agreements at regular intervals. Rewrite and renegotiate them to assure understanding on both sides.

11. Assuming suppliers provide products that meet necessary specifications. *Remedy*—have set specifications and inspect supplies received for conformance.

12. Assuming that there is no need to change designs or procedures if there are no complaints. *Remedy*—this may be too late under present legislation. Check things before shipping.

13. Assuming a safety program is not needed because there is insurance coverage. *Remedy*—not true under present legislation. Insurance does not cover all costs and penalties.

It would be possible to list other misconceptions and assumptions of those employers and managers who tend to operate "off the cuff." The business, social, economic, and political climate will no longer allow this type of operation. Employers and managers must become ever more professional and thorough in their jobs.

Setting Up an Effective Safety and Health Program

When a high degree of safety is incorporated into the design of the equipment or the planning of the process, the need for training and supervision to control unsafe acts is reduced.

The most effective time to keep hazards out of the plant, product, process, or job is *(a)* prior to building or remodeling, *(b)* while a product is being designed, *(c)* before a change in a process is put into effect, or *(d)* before a job is started. Every effort, therefore, should be made to find and remove potential hazards at the blueprint or planning stage.

In most accidents, both an unsafe condition and an unsafe act are contributing factors. Unsafe conditions often can lead employees to perform unsafe acts. For example, an unsafe act may be caused by poor machine design, inadequately planned methods, and other engineering deficiencies. Thus, elimination of a hazard caused by an unsafe condition may also reduce the likelihood of injury from an unsafe act.

When safety is being designed into an operation, consideration must be given to human factors.

18

Almost every machine or piece of equipment needs an operator so the design must consider human limitations and capabilities, both mental and physical. Human factors engineering, ergonomics, biomechanics, and psychophysics are some other names for this discipline. Human factors engineering is concerned with the interaction (or interface) between man and machine, and man and environment.

The success of any safety and industrial hygiene and health program requires the cooperative effort of the engineering, safety, medical, industrial hygiene, and purchasing organizations, and the supervisor and the individual employee.

There are many firms and businesses that do not have an individual for each of these functions. They may have one individual who performs more than one function, or they may contract for help as needed, or both.

Engineering function

The introduction of new processes and operations begins with engineering. For this reason, the engineer plays a most important role in the control of occupational safety and health hazards.

The engineer's responsibilities are:

To plan all operations using established engineering procedures to prevent unnecessary exposure to harmful environmental factors or stresses.

To check the medical, industrial hygiene, and safety aspects whenever new operations or processes are planned.

To request a safety and industrial hygiene survey of new installations before permitting shop personnel to operate the equipment.

Safety function

The safety professional plays an integral part in the overall environmental health program. His responsibilities are:

To conduct an effective safety program by coordinating the educational, engineering, supervisory, and enforcement activities related to it.

To work with hygienists and health professionals to make sure that employees are provided with a healthful work environment.

To provide educational material for any safety training program for personnel working with hazardous materials or processes.

To assist the supervisor in teaching his employees safety rules, regulations, and procedures.

To conduct safety surveys to make sure that proper practices and procedures are being followed.

To recommend changes in safety rules, regulations, and procedures to keep pace with technological advancements.

Industrial hygiene function

The industrial hygienist is primarily responsible for monitoring the work environment and assisting the process engineer in making sure that a safe environment is provided and maintained. He has a responsibility:

To advise appropriate organiza-

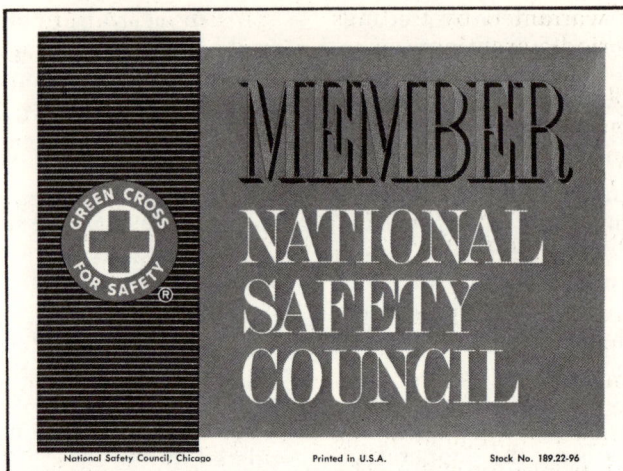

Figure 1–1.—A sticker like this one (actual size $4^{1/4}$ by $5^{1/2}$ in.) can be placed on bulletin boards, entrance doors, and in employment offices to remind everyone of a company's continuous efforts to remove hazards and prevent accidents.

tions of the potential hazard arising out of any current or proposed process or operation.

To establish hygiene standards, and to make appropriate tests periodically to make sure that the standards are met.

To specify the design and quality of all types of personal protective equipment and to prescribe standards for their use.

To recommend controls necessary to minimize employee exposure to harmful environmental factors or stresses.

To assist the supervisor in educating employees on practices, precautions, and procedures established to control their occupational exposure.

To review present and proposed practices to make sure that they are in accordance with estab-

lished standards.

To survey all new installations before they are turned over to shop personnel.

Medical function

The responsibilities of the physician in the safety and health program are:

To recommend the placement only of those employees whose physical and emotional health capacities meet the minimum job requirements.

To cooperate in the development of adequate and effective measures to prevent exposure to harmful agents.

To examine periodically those employees who work with or are exposed to hazardous materials.

To restrict employees from further exposure on a medical basis

20

whenever warranted by findings of such periodic examinations.

Purchasing function

The purchasing department should purchase only equipment and materials approved by the industrial hygiene, safety, medical, and other appropriate departments.

Supervisor's responsibilities

Each supervisor is responsible for maintaining safe working conditions within his organization and for directly implementing the safety and health program. His responsibilities are:

To maintain a work environment that assures the maximum safety and health for his employees.

To make certain that applicants (newly hired or transferred) have been medically examined and approved for the job before being assigned to work.

To instruct employees periodically on precautions, procedures, and practices to be followed to eliminate accidental exposure to hazards to safety and health.

To make sure that meticulous housekeeping practices are developed and employed at all times.

To make sure that food, candy, beverages, other edibles, and tobacco are not stored or consumed in work areas where toxic materials may be present.

To inform promptly the engineering, industrial hygiene, and safety personnel of any operation or condition which appears to present a hazard to employees.

To inform medical personnel promptly in case of accidental exposure, and to send the employee(s) involved for a medical examination.

To furnish his employees with the proper personal protective equipment, instruct them in its proper use, and enforce the wearing of such equipment.

To observe all work restrictions imposed by the medical organization.

To consult with safety, industrial hygiene, engineering, and medical personnel for aid in fulfilling his responsibilities.

To administer appropriate disciplinary action when safety rules are violated.

Employee's responsibilities

Each employee is responsible for contributing his part towards the success of the safety and health program. His part is:

To notify his supervisor immediately when certain conditions or practices may cause personal injury or illness or property damage.

To observe all safety rules and to make maximum use of all prescribed personal protective equipment and to follow practices and procedures established to maintain his health and safety.

To report any accident immediately to his supervisor.

To develop and practice good habits of personal hygiene and housekeeping.

OSHA Considerations

A new national policy was established on December 29, 1970, when President Richard M. Nixon signed into law the Occupational Safety and Health Act of 1970, usually referred to as the OSHAct. The Congress of the United States declared that the purpose of this piece of legislation is "to assure so far as possible every working man and woman in the Nation safe and healthful working conditions and to preserve our human resources. . . ."

Who Is Covered by OSHA

The OSHAct, which took effect April 28, 1971, covers approximately 60 million employees throughout the Nation.

Except for specific exclusions, the Act is applicable to every employer who has one or more employees and who is engaged in a business affecting commerce. The law applies to all 50 states, the District of Columbia, Puerto Rico, and all U.S. possessions.

Specifically excluded from coverage are all federal, state, and local government employees. There are, however, special provisions in the Act for federal employees and potential coverage for state and local government employees.

The OSHAct is also not applicable to those operations where a federal agency (and state agencies acting under the Atomic Energy Act of 1954) other than the Department of Labor has statutory authority to prescribe or enforce standards or regulations affecting occupational safety or health. An example of this exclusion are in the operations covered by the federal Metal and Nonmetallic Mine Safety Act or the federal Coal Mine Health and Safety Act, where statutory authority is vested in the Department of the Interior.

Churches and religious organizations, with respect to their religious activities, are not regarded as employers. Likewise, persons who in their own residences employ others to perform domestic household tasks are not regarded as employers. Further, any person engaged in agriculture who is a member of the immediate family of the farmer is not regarded as an employee and hence is not covered by the Act.

Employer duties

Each employer covered by the Act:

- Has the general duty to furnish each of his employees employment and a place of employment which is free from recognized hazards that are causing or likely to cause death or serious physical harm (this is commonly known as the "general duty clause"); and
- Has the specific duty of complying with safety and health standards promulgated under the Act. (See the next section for details.)

For employers, the general duty provision is used only where there are no specific standards applicable to a particular hazard involved. A hazard is "recognized" if it is a condition that is generally recognized as a hazard in the particular industry in which it occurs and is detectable (a) by means of the human senses, or (b) by means of accepted tests known in the industry to determine its existence which should make its presence known to the employer. An example of a "recognized hazard" in the latter category is excessive concentrations of a toxic substance in the work area atmosphere, even though such concentrations can only be detected through use of measuring devices.

Employee duties and rights

Each employee, in turn, has the duty to comply with the safety and health standards and all rules, regulations and orders which are applicable to his own actions and conduct on the job.

However, there is no provision for government sanctions against an employee for failure to comply.

While the law expressly places upon each employee the obligation to comply with the standards, final responsibility for compliance with the requirements of the Act remains with the employer. Employers, therefore, should take all necessary action to assure employee compliance with the standards and establish within their safety program a means of finding out when employees are not complying with applicable standards.

While the employee has the legal duty to comply with all the standards and regulations issued under the OSHAct, there are many employee rights that are also incorporated in the Act. Since these rights may affect labor relations as well as labor negotiations, employers as well as employees should be aware of the employee rights contained in the Act. Employee rights fall into three main areas and are related to (a) standards, (b) access to information, and (c) enforcement.

Rights with respect to standards.

1. Employees may request the Occupational Safety and Health Administration (OSHA) to begin proceedings for adoption of a new standard or to amend or revoke an existing one.

2. Employees may submit written data or comments on proposed standards and may appear as an interested party at any hearings held by OSHA.

3. Employees may file written objections to a proposed federal standard and/or appeal the final decision of OSHA.
4. Employees must be informed when an employer applies for a variance of a promulgated standard.
5. Employees must be afforded the opportunity to participate in a variance hearing as an interested party and have the right to appeal OSHA's final decision.

With respect to access to information.
1. Employees have the right to information from the employer regarding employee protection and obligations under the Act.
2. Affected employees have a right to information from the employer regarding the toxic effects, conditions of exposure, and precautions for safe use of all hazardous materials in the establishment by means of labeling or other forms of warning where such information is prescribed by a standard.
3. If employees are exposed to harmful materials in excess of levels set by the standards, the affected employees must be so informed by the employer and the employer must also inform the employees thus exposed what corrective action is being taken.
4. If a compliance officer (CO) determines that an alleged imminent danger exists, he must inform the affected employees of the danger and that he is recommending that relief be

sought by court action if the imminence of such danger is not eliminated.
5. Upon request, employees must be given access to records of their history of exposure to toxic materials, or harmful physical agents which are required to be monitored or measured and recorded.
6. If a standard requires monitoring or measuring hazardous materials or harmful physical agents, employees must be given the opportunity to observe such monitoring or measuring.
7. Employees have the right of access to *(a)* the list of toxic materials published by National Institute for Occupational Safety and Health (NIOSH), *(b)* criteria developed by NIOSH describing the effects of toxic materials or harmful physical agents, and *(c)* industry-wide studies conducted by NIOSH regarding the effects of chronic, low-level exposure to hazardous materials.
8. On written request to NIOSH, employees have the right to obtain the determination of whether or not a substance found or used in the establishment is harmful.

With respect to enforcement.
1. Employees have the right to confer with the compliance officer in connection with an inspection of an establishment.
2. An authorized employee representative must be given an

opportunity to accompany the compliance officer during an inspection for the purpose of aiding such inspection. (This is commonly known as the "walk-around" provision.)

3. An employee has the right to make a written request to OSHA for a special inspection if the employee believes a violation of a standard threatens physical harm, and the employee has the right to request OSHA to keep his identify confidential.

4. If an employee believes any violation of the Act exists, he has the right to notify OSHA or a compliance officer in writing of the alleged violation, either before or during an inspection of the establishment.

5. If a request is made for a special inspection and it is denied by OSHA, the employee must be notified in writing by OSHA, together with the reasons that the complaint was not valid. The employee has the right to object to such a decision and may request a hearing by OSHA.

6. If a written complaint concerning an alleged violation is submitted to OSHA and the compliance officer responding to the complaint fails to cite the employer for the alleged violation, OSHA must furnish the employee or his authorized representative a written statement setting forth the reasons for its final disposition.

7. If OSHA cites an employer for a violation, employees have the right to review a copy of the citation which must be posted by the employer at or near the place where the violation occurred.

8. Employees have the right to appear as an interested party or to be called as a witness in a contested enforcement matter before the Occupational Safety and Health Review Commission.

9. If OSHA arbitarily or capriciously fails to seek relief to counteract an imminent danger and an employee is injured as a result, the employee has the right to bring action against OSHA for relief as may be appropriate.

10. An employee has the right to file a complaint to OSHA within 30 days if he believes he has been discriminated against because he asserted his rights under the Act.

11. An employee has the right to object to the abatement period fixed in the citation issued to his employer by appealing to the Occupational Safety and Health Review Commission.

Who Administers the Act

Administration and enforcement of the OSHAct are vested primarily with the Secretary of Labor and the Occupational Safety and Health Review Commission. With respect to the enforcement function, the Secretary of Labor performs the investigation and prosecution aspects of the

25

enforcement process and the Review Commission performs the adjudication portion of the enforcement process.

Research and related functions and certain educational functions are vested in the Secretary of Health, Education, and Welfare and are, for the most part, carried out by the National Institute for Occupational Safety and Health (NIOSH) established within the Department of Health, Education, and Welfare (HEW). Compiling injury and illness statistical data is handled by the Bureau of Labor Statistics, U.S.Department of Labor.

Occupational Safety and Health Administration

The Occupational Safety and Health Administration (OSHA) is divided into three major program areas. These program areas and their major functions are:

National programs.
1. Develop and promulgate occupational safety and health standards and issue regulations.
2. Provide training programs for OSHA personnel as well as employers and employees in an effort to improve unsafe and unhealthful working conditions.
3. Assist in establishing federal agency plans and programs to assure effective compliance to standards within the federal agencies.

Regional programs.
1. Provide for national support in development of state compli-

ance programs and approve same.
2. Assure the execution of an effective compliance system through inspection and technical assistance.
3. Review and assess the effectiveness and efficiency of federal/state compliance activities.

Administrative programs.
1. Provide financial, personnel, and administrative services for OSHA.
2. Perform management reviews and analysis for improvement of services for OSHA.
3. Provide for data collection and analysis, and design of information systems to facilitate the management of OSHA programs.

To assist in carrying out its responsibilities, OSHA has established ten regional offices. The primary mission of the regional office is to supervise, coordinate, evaluate, and execute all programs of OSHA in the region.

Area offices have been established within each region, each headed by an area director. The mission of the area director is to carry out the compliance program of OSHA within designated geographic areas. The real action for implementing the enforcement portion of the OSHAct is carried out by the area offices.

Occupational Safety and Health Review Commission

The Occupational Safety and Health Review Commission (OSHRC) is a quasi-judicial board of three members appointed by the President and confirmed by the Senate. The Commission is an independent agency of the Executive Branch of the U.S. Government and is not a part of the Department of Labor. The principal function of the Commission is to adjudicate cases resulting from an enforcement action initiated against an employer by OSHA when any such action is contested by the employer or by his employees or their representatives.

The Commission's actions are limited to contested cases. In such cases, OSHA notifies the Commission of the contested cases and the Commission hears all appeals on actions taken by OSHA concerning citations, proposed penalties and abatement periods, and determines the appropriateness of such actions. When necessary, the Commission may conduct its own investigation and may affirm, modify, or vacate OSHA's findings.

There are two levels of adjudication within the Commission: (a) the administrative law judge, and (b) the three-member Commission. All cases not resolved in informal proceedings are heard and decided by one of the Commission's administrative law judges. The judge's decision can be changed by a majority vote of the Commission if one of the members, within 30 days of the judge's decision, directs that the judge's decision be reviewed by the Commission members. The Commission is the final administrative authority to rule on a particular case, but its findings and orders can be subject to further review by the courts.

National Institute for Occupational Safety and Health

The National Institute for Occupational Safety and Health (NIOSH) was established within the HEW under the provisions of the OSHAct. Administratively, NIOSH is located in HEW's Center for Disease Control. NIOSH is the principal federal agency engaged in research, education, and training related to occupational safety and health.

The primary functions of NIOSH are to (a) develop and establish recommended occupational safety and health standards, (b) conduct research experiments and demonstrations related to occupational safety and health, and (c) conduct education programs to provide an adequate supply of qualified personnel to carry out the purposes of the OSHAct.

Research and related functions. Under the OSHAct, NIOSH has the responsibility for conducting research for new occupational safety and health standards. NIOSH develops criteria for the establishment of such standards. Such criteria are transmitted to OSHA which has the responsibility for the final setting, promulgation, and enforcement of the standards.

The OSHAct also requires NIOSH to publish an annual listing of all known toxic substances and the concentrations at which such toxicity is known to occur. While the inclusion of a substance on the list does not mean that it is to be avoided, it does mean that the listed substance has a documented potential of being hazardous if misused and, therefore, care must be exercised to control the substance. Conversely, the absence of a substance from the list does not necessarily mean that a substance is nontoxic. Some hazardous substances may not qualify to be listed because the dose that causes the toxic effect is not known.*

Education and training. NIOSH also has the responsibility to conduct *(a)* education and training programs which are aimed at providing an adequate supply of qualified personnel to carry out the purpose of the Act, and *(b)* informational programs on the importance and proper use of adequate safety and health equipment. The long-term approach to an adequate supply of training personnel in occupational safety and health is found in the colleges and universities and private institutions. NIOSH encourages them by contracts and grants, to expand their curricula in occupational medicine, occupational health nursing, industrial hygiene, and occupational safety engineering.

Employer and employee services. Of principal interest to individual employers and employees are the technical services offered by NIOSH. Five main services are provided upon request to NIOSH's Division of Technical Services, Cincinnati, Ohio 45202.

1. *Hazard evaluations.* Provides an on-site evaluation of potentially toxic substances used or found on the job. This must be done by request using Form NIOSH-4(Cin)— see Figure 2-1.
2. *Technical information.* Provides technical information concerning health or safety conditions at workplaces such as the possible hazards of working with specific solvents, and when to use protective equipment.
3. *Accident prevention.* Provides technical assistance for controlling on-the-job injuries including the evaluation of special problems and recommendations for corrective action.
4. *Industrial hygiene.* Provides technical assistance in the areas of engineering and industrial hygiene, including the evaluation of special health-related problems in the workplace and recommendations for control measures.
5. *Medical services.* Provides assistance in solving occupational medical and nursing problems in the workplace including the assessment of existing medically related needs and the development of recommended means for meeting such needs.

---

*The *Toxic Substance List* (Annual Edition) is available and for sale by the Superintendent of Documents, U.S. Government Printing Office, Washington, D.C. 20402.

U.S. DEPARTMENT OF HEALTH, EDUCATION, AND WELFARE
NATIONAL INSTITUTE FOR OCCUPATIONAL SAFETY AND HEALTH

REQUEST FOR HEALTH HAZARD EVALUATION

This form is provided to assist in registering a request for a health hazard evaluation with the U.S. Department of Health, Education, and Welfare as provided in Section 20(a)(6) of the Occupational Safety and Health Act of 1970 and 42 CFR Part 85. (See Statement of Authority on Reverse Side).

Name of Establishment Where Alleged Hazard(s) Exist _____

Company { Street _____ Telephone _____
Address { City _____ State _____ Zip Code _____

1. Principal Company Activity _____
   (manufacturing, construction, transportation, services, etc.)
2. Specify the particular building or worksite where the alleged hazard is located, including address _____
   _____

3. Specify the name and phone number of employer's agent(s) in charge.
   _____

4. Describe briefly the hazard(s) which exists by completing the following information:
   Identification of Hazard or Toxic Substance(s)_____
   Trade Name (If Applicable) _____Chemical Name _____
   Manufacturer _____ Does the material have a warning label? _____ Yes _____ No
   If Yes, attach copy of label or a copy of the information contained on the label.
   Physical Form: Dust ☐   Gas ☐   Liquid ☐   Mist ☐   Other ☐
   Type of Exposure?   Breathing ☐   Swallowing ☐   Skin Contact ☐
   Number of People Exposed _____ Length of Exposure (Hours/Day) _____
   Occupations of Exposed Employees _____
   _____
   _____

5. Using the space below describe further the nature of the conditions or circumstances which prompted this request and other relevant aspects which you may consider important, such as the nature of the illness or symptoms of exposure, the concern for the potentially toxic effects of a new chemical substance introduced into the
   workplace, etc.
   _____
   _____
   _____
   _____
   _____

NIOSH-4 (Cin)   FORM APPROVED
9/72   OMB NO. 68-R1236

Figure 2–1.—NIOSH will evaluate hazards due to toxic substances.

Bureau of Labor Statistics

The responsibility for conducting statistical surveys and establishing methods used to acquire injury and illness data is placed in the Bureau of Labor Statistics (BLS). Questions regarding recordkeeping requirements and reporting procedures can be directed to any of the OSHA regional or area offices or the BLS regional offices. Details are given in Chapter 5, Accident Records and Reports.

Federal-state relationships

The OSHAct encourages the states to assume the fullest responsibility for the administration and enforcement of their own occupational safety and health laws. Any state may assume responsibility for the development and enforcement of occupational safety and health standards relating to any occupational safety and health issue covered by a standard promulgated under the OSHAct. However, in order to assume this responsibility, such state must submit a state plan to OSHA for approval. When such a plan satisfies designated conditions and criteria, OSHA approves the plan.

Following approval of a state plan, OSHA will continue to exercise its enforcement authority until it determines on the basis of actual operations that the state plan is indeed being satisfactorily carried out. If the implementation of the state plan is satisfactory during the first three years after its approval, then the federal standards and federal enforcement of such standards under the OSHAct can become inapplicable with respect to issues covered under the plan. This means that for the interim period of dual jurisdiction, employers must comply with the state standards as well as the federal standards. Should the state agency responsible fail to fully implement the state plan and the state's performance fails to be "at least as effective as" the federal program, OSHA has the obligation to withdraw its approval of the state plan and once again assume full jurisdiction in that state.

Safety and Health Requirements

The Act authorizes OSHA to promulgate, modify, or revoke occupational safety and health standards.

In order to get the initial set of standards in place without undue delay, the Act authorized OSHA to promulgate an existing federal standard or any national consensus standard without regard to the usual rulemaking procedures prior to April 28, 1973. The standards appear in Title 29—Labor, *Code of Federal Regulations,* Chapter XVII—Occupational Safety and Health Administration, and are continually updated.*

---

*Copies of all OSHA standards are available at OSHA Regional and Area offices, and are for sale by the Superintendent of Documents, U.S. Government Printing Office, Washington, D.C. 20402. The *Code of Federal Regulations* and the *Federal Register* are also for sale by the U.S. Government Printing office.

Standards contained in Part 1910 are applicable to general industry. Those contained in Part 1926 are applicable to the construction industry. Standards applicable to ship repairing, shipbuilding, shipbreaking and longshoring are contained in Parts 1915 through 1918, respectively.

OSHA also has the authority to promulgate emergency temporary standards where it is found that employees are exposed to grave danger. Emergency temporary standards can take effect immediately upon publication in the *Federal Register*. Such standards cannot remain in effect, but do until superceded by a standard promulgated under the procedures prescribed by the Act.

Any person adversely affected by any standard issued by OSHA has the right to challenge its validity by petitioning the U.S. Court of Appeals within 60 days after its promulgation.

In order to promulgate, revise, or modify a standard, OSHA must first publish in the *Federal Register* a notice of any proposed rule that will adopt, modify, or revoke any standard and invite interested persons to submit their views on the proposed rule within 30 days after publication. Interested persons may file objections to the rule and are entitled to a hearing on their objections if they request a hearing be held. However, objections must specify the parts of the proposed rule to which they object and the grounds for such objection. If a hearing is requested, OSHA must hold one. Based on *(a)* the need for control of an exposure to an occupational injury or illness, and *(b)* the reasonableness, effectiveness, and feasibility of the control measures required, OSHA may issue a rule promulgating an additional standard or modify or revoke an existing standard.

Variances from standards

There will be some occasions when, for various reasons, standards cannot be met by some employers. In other cases, the protection already afforded by an employer to employees is equal to or superior to the protection that would be granted if the standards were followed strictly to the letter. The Act provides an avenue of relief from these situations by empowering OSHA to grant variances from the standards which would not degrade the purpose of the Act.

There are two types of variances—temporary and permanent. An employer may apply for an order granting a temporary variance provided he establishes that *(a)* he cannot comply with the applicable standard because of unavailability of personnel or equipment or time to construct or alter facilities, *(b)* he is taking all available steps to protect his employees against exposure covered by the standard, *(c)* his program will effect compliance with the standard as soon as possible, and *(d)* he certifies that he has informed his employees of the application and of their rights to petition for a hearing.

An employer may also apply for a permanent variance from a standard. A variance order can be granted if OSHA determines that an employer has demonstrated by a preponderance of evidence that he will provide a place of employment as safe and healthful as that which would prevail if he complied with the standard.

An employer may request an interim order permitting either kind of variance until his formal application can be acted upon. Again, the request for an interim order must contain statements of fact or arguments why such interim order should be granted. If a request for an interim order is denied, the applicant will be notified promptly and informed of the reasons for the decision. If the order is granted, all concerned parties will be informed and the terms of the order will be published in the *Federal Register.* In such cases, the employer must inform the affected employees regarding the interim order in the same manner used to inform them of the variance application.

Upon filing an application for a variance, OSHA will publish a notice of such filing in the *Federal Register* and invite written data, views, and arguments regarding the application. Those affected by the petition may request a hearing. After reviewing all the facts, including those presented during the hearing, OSHA publishes its decision regarding the application in the *Federal Register.*

Recordkeeping and reporting requirements

Most employers covered by the Act are required to maintain, in each establishment, records of recordable occupational injuries and illness. The regulations pertaining to recordkeeping and reporting of occupational injuries and illnesses are promulgated in 29 CFR 1904. They are described in Chapter 5, Accident Records and Reports.

How To Handle Enforcement Actions

The OSHA poster

The OSHA poster (see Figure 2-2) must be prominently displayed in a conspicuous place in the workplace where notices to employees are customarily posted. The poster informs employees of their rights and responsibilities under the Act.

Inspection procedures

The Department of Labor compliance officer (CO) may enter, at any reasonable hour, and without delay, any establishment covered by the OSHAct to inspect the premises and all its facilities. CO's also have the right to question privately any employer, owner, operator, agent, or employee.

The OSHAct authorizes an employee representative to accompany the CO during the official inspection of the premises and all its facilities. Usually, the authorized employee representative will be the union steward or the chairman of the employee safety committee. Occasionally there may be no authorized employee representative, especially in those establishments which are nonunion shops. In the absence of an employee representative, the CO will confer with employees picked at random.

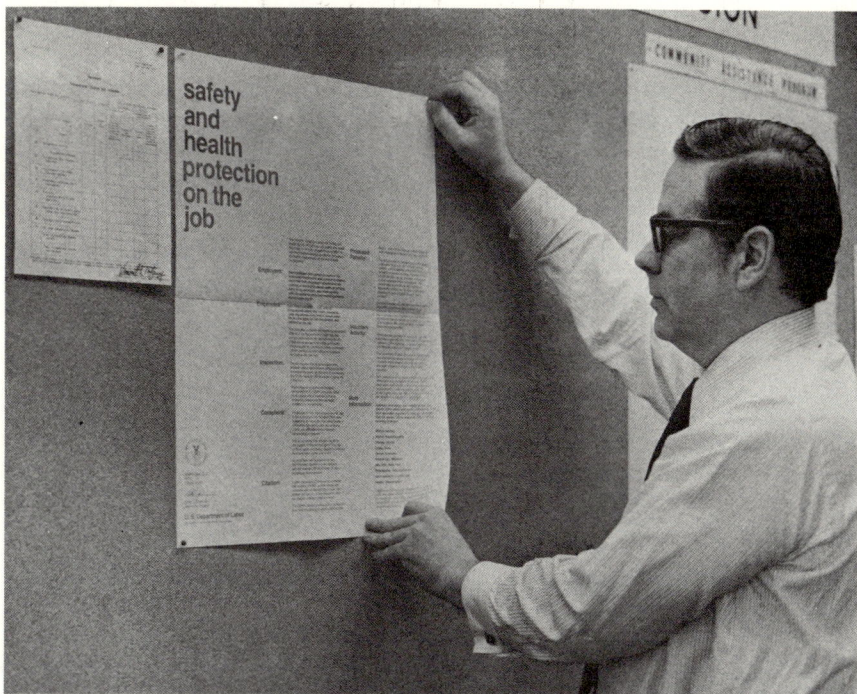

Figure 2–2.—OSHA poster, "Safety and Health Protection on the Job," must be posted conspicuously at every plant, job site, or other establishment. At left is the annual Summary of Occupational Injuries and Illnesses, OSHA Form 102, that must be posted by February 1 of the following year and remain in place for the next 30 consecutive days.

Representatives of the Department of Health, Education, and Welfare, although not authorized to enforce the OSHAct, are authorized to make inspections and to question employers and employees in order to carry out those duties assigned to HEW under the Act.

Inspection priorities. OSHA has established priorities for assignment of manpower and resources. Such priorities are as follows:

1. The elimination of imminent-danger situations—usually brought to OSHA's attention through external communications (such as employee complaints).

2. Investigation of a fatality or catastrophic event (an event that results in the hospitalization of five or more employees).

3. Response to valid complaints from employees.

4. Target Industry Program and Target Health Hazards Program.

The industries that were initially included in the Target Industry

33

Program are: longshoring, lumber and wood products, roofing and sheet metal, meat and meat products, and miscellaneous transportation (manufacture of mobile homes and other transportation equipment). The toxic substances that were initially included in the Target Health Hazards Program are: asbestos, silica, lead, cotton dust, and carbon monoxide.

5. General inspections and related activities to provide broad representative coverage. Random selection factors include geographical distribution, size distribution, and cross-section of industry.

The General inspection procedures. The primary responsibility of the CO (Compliance officer), who is under the supervision of the OSHA area director, is to conduct an effective inspection to determine if employers and employees are in compliance with the requirements of the standards, rules, and regulations promulgated under the OSHAct.

To enter an establishment, the CO will present his credentials to a guard, receptionist, or other person acting in such a capacity. There have been instances where some credentials have been fraudulent. Employers, therefore, should check the CO's credentials carefully before letting the individual enter their establishment for the purpose of an inspection. (See Figure 2-3.)

The CO will usually ask for the person with the highest authority that is on the premises. It is recommended that employers furnish written instructions to the security, receptionist, or other affected personnel regarding the right of entry, treatment, who should be notified, and to whom and where the CO should be directed so that undue delay can be avoided.

Opening conference

The employer or his designated representative will be required to participate in the opening conference conducted by the CO. Since the CO will want to talk with safety personnel, if any, such personnel should also be invited by the employer to the opening conference. The employee representative who accompanies the CO through the establishment should also participate in the opening conference.

At the opening conference, the CO will:

1. Inform the employer that the purpose of his visit is to make an investigation to ascertain if the establishment, procedures, operations, and equipment are in compliance with the requirements of the OSHAct.

2. Give the employer copies of the Act, standards, regulations, and promotional material, as necessary.

3. Outline in general terms the scope of the inspection, the records he wants to review, and his obligation to confer with employees.

4. If applicable, furnish a copy of any complaint(s).

5. Answer questions those in attendance might have.

During the course of the opening conference the CO may want

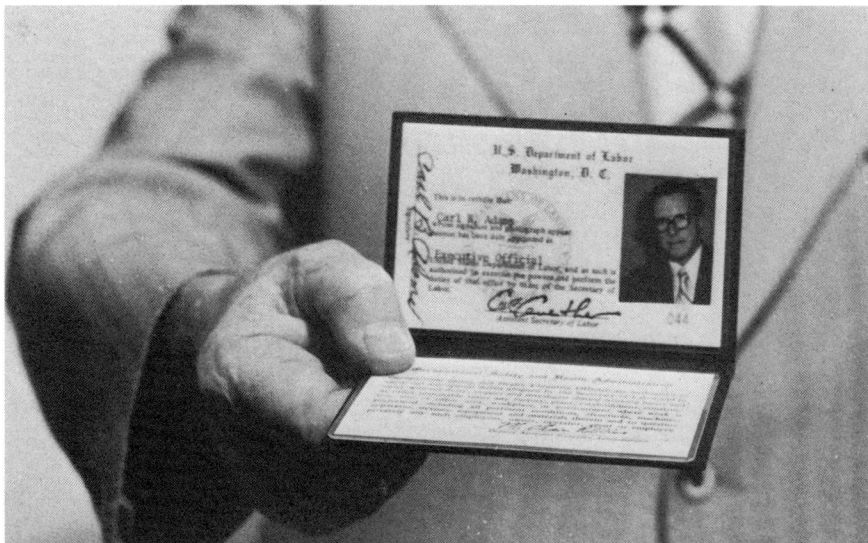

Figure 2–3.—Bona-fide OSHA compliance officers are equipped with official identification as shown here. The credentials are signed by an assistant secretary of labor. If in doubt about the validity of the credentials, it is recommended that the employer contact the nearest OSHA area office and determine whether or not the area office has scheduled an inspection at the establishment in question.

to review OSHA Form 100, Log of Occupational Injuries and Illnesses, and OSHA Form 101, Supplementary Record of Occupational Injuries and Illnesses; described in Chapter 5, Accident Records and Reports. Such records should be made readily available to him. In addition, the CO will want to obtain information regarding the safety and health program that the employer now has in effect so that he can evaluate such a program. Naturally, a comprehensive safety and health program that shows evidence of effective performance in accident prevention will be impressive to all concerned.

The CO will also ascertain from the employer whether employees of another employer (for example, a contracting employer for maintenance or remodeling) are working in or on the establishment. If so, the CO will afford the authorized representative of those employees a reasonable opportunity to accompany him during the inspection of the workplaces where they are working.

During the conference the CO will explain the employee representative's rights and ask for the authorized employee representative. Generally the employee representative will be an employee of the establishment inspected. However, if in the judgement of the CO, good cause has been

35

shown that accompaniment of a third party (such as an industrial hygienist or safety consultant) who is not an employee of the employer (but is indeed, an authorized employee representative) is reasonably necessary to conduct an effective and thorough inspection, such a third party may accompany the CO during the inspection. The final decision will rest with the CO.

The employer is not permitted to designate the employee representative. Employee representatives may change as the inspection process moves from job location to job location. The CO may deny the right of accompaniment to any person whose conduct interferes with a full and orderly inspection. If there is no authorized employee representative, the CO will consult with a reasonable number of employees concerning matters of safety and health in the workplace during the course of the inspection.

Inspection of facilities

The CO will take the time necessary to inspect all of the operations in the establishment. The inspectors have as their primary objective the enforcement of the occupational safety and health standards as well as the enforcement of other promulgated regulations such as the posting of the OSHA poster, Figure 2-2, page 33.

The CO will have the necessary instruments for checking certain items, such as noise levels, certain air contaminants and toxic substances, electrical grounding, and the like. During the course of inspection, the CO will note any apparent violation of the standards, and will normally record any apparent violation, including its location, and any comments that he has regarding the violation. He will do the same for any apparent violation of the general duty clause. His notes will serve as the basis of information for the area director for issuing citations or proposed penalties. For these reasons, the employer representative should ascertain any apparent violations from the CO during the actual inspection of the facilities. The employer representative should make notes identical to the CO's during the actual inspection so that he will have precisely the same information that the CO has.

It should be noted that the CO is only required to record apparent violations and is not required to present a solution or method of correcting, minimizing, or eliminating the violation. OSHA, however, will respond to requests for technical information concerned with complying with given standards. In such cases, the employer is urged to contact the regional or area office.

If, during the course of an inspection, the CO receives a complaint from an employee regarding a condition which is alleged to be in violation of an applicable standard, the CO, even though the complaint is brought to him via an informal process, will normally inspect for the alleged violation.

In the couse of his normal inspection, the CO may make some preliminary judgments with respect to environmental conditions affecting occupational health. In such cases, he will generally use direct-reading instruments. Should this occur, and if proper instrumentation is available, it would be prudent for the employer to have qualified personnel at the establishment make duplicate tests in the same area at the same time under the same conditions. In addition, the employer representative should again take careful notes on the CO's methods as well as the results. If the inspection indicates a need for an indepth industrial hygiene analysis, the CO will notify the OSHA area director, who may assign a qualified industrial hygienist to investigate further. If a laboratory analysis is required, samples are sent to OSHA's laboratory in Salt Lake City and the results will be reported back to the area director.

At the completion of the inspection, but prior to the beginning of the closing conference, the employee representative is usually excused.

Closing conference

Upon completion of the inspection, the CO will confer with the employer or his representative. Again, the employer's safety personnel, if any, should be present at the closing conference. It is at this time that the CO will advise the employer and his representatives of all conditions and practices which may constitute a safety or health violation. He should also indicate the applicable section or sections of the standards which may have been violated.

The CO will normally advise the employer that citations may be issued for alleged violations and that penalties may be proposed for each violation. Administratively, the authority for issuing citations and proposed penalties rests with the area director or his representative.

The employer will also be informed that citations will fix a reasonable time for abatement of the violations alleged. The CO will attempt to obtain from the employer an estimate of the time he feels would be required to abate the alleged violation and take such estimate into consideration when recommending a time for abatement. The CO should explain the appeal procedures with respect to any citation or any notice of a proposed penalty.

Followup inspections

A followup inspection will always be made for those situations involving imminent danger or where citations have been issued for serious, repeated, or willful violations. Followup inspections for all other cases will be conducted at the discretion of the area director.

The followup inspection is intended to be limited to verifying compliance of those conditions which were alleged to be in violation. The followup inspection is conducted with all of the usual formality of the original inspection, including the opening and closing conferences, and the walk-around rights of the employer and employee representative.

Alleged Violations

In addition to the general duty clause, the occupational safety and health standards promulgated under the OSHAct are used as a basis for determining alleged violations. There are four types of violations: imminent danger, serious, nonserious, and de minimis (very minor).

Imminent danger

The OSHAct defines imminent danger as "any conditions or practices in any place of employment which could reasonably be expected to cause death or serious physical harm immediately or before the imminence of such danger can be eliminated through the enforcement procedures otherwise provided by this Act." Therefore, for conditions or practices to constitute an imminent danger situation, there must be a reasonable certainty that immediately or within a short period of time such conditions or practices could result in death or serious physical harm. Normally a health hazard would not constitute an imminent danger except in *extreme* situations, such as the presence of potentially lethal concentrations of airborne toxic substances that are an immediate threat to the life or health of employees.

If, during the course of inspection, the CO deems that the existing set of conditions appears to constitute an imminent danger situation, he will immediately advise the employer or his representative that such a danger exists and will attempt to have the danger corrected immediately through voluntary compliance. Further, if any employees appear to be in imminent danger, they will be informed of the danger and the employer will be requested to remove them from the area of imminent danger.

An employer will be deemed to have abated the imminent danger by *(a)* removing employees from the danger area, or *(b)* eliminating the conditions or practices which constitute the imminent danger. Normally abatement is achieved in either of these two ways. If the employer refuses to voluntarily abate the alleged imminent danger, the CO will inform the affected employees of the danger involved and will inform the employer as well as the affected employees that he will recommend a civil action (in the form of a court order) for appropriate relief (for example, to shut down the operation). In such cases, the CO will personally post the imminent danger citation at or near the area in which the exposed employees are working. The federal CO has no authority to order the closing down of an operation or to direct employees to leave the area of imminent danger or the workplace.

Serious violation

To determine if a violation is serious, the CO must decide:

• Is there a substantial probability that death or serious physical harm could result? And if so,

• Did the employer know, or with the exercise of reasonable diligence, should he have known of the hazard?

If the answer to both questions is "yes," then a serious violation exists.

The term "serious physical harm" is similar to the "permanent total disability" or "permanent partial disability" utilized in the ANSI Z16.1, *Method of Recording and Measuring Work Injury Experience.* These are discussed in Chapter 5, Accident Records and Reports.

Serious physical harm includes impairment of the body where:

• A part of the body would be permanently removed or rendered functionally useless (such as an amputation or loss of an eye) or substantially reduced in efficiency on or off the job.

• A part of an internal bodily system would be inhibited in its normal performance to such a degree as to shorten life or cause reduction in physical or mental efficiency (such as severe loss of hearing).

It is obvious that the CO must make an evaluation as to the *likelihood* that death or serious physical harm could result from a condition which is an alleged violation. For example, a violation involving an inadequate guard rail for workers at a substantially high level from the ground would normally be a serious violation since it is more probable than not that the result of such a condition would be death or serious physical harm.

Note that the emphasis in deciding whether or not a condition represents a serious violation is based on the *seriousness* or *severity* of the most likely injury that would arise out of the potential accident, rather than on the *prob-ability* that the accident will occur as a result of the violation. In all cases, the decision in determining whether a violation is serious or not will require professional judgment.

Nonserious violation

If the more likely consequence of a violation is something less than death or serious physical harm, or the employer did not know of the hazard, then the violation will be considered a nonserious violation. For example, a violation of housekeeping standards that might result in a tripping accident would be classified as a nonserious violation since the more probable consequence of such a condition would be a strain or contusion, neither of which is classified as serious physical harm.

De minimis violations

De minimis violations are those that have no immediate or direct relationship to safety or health.

Special types

There are two other special types of violations: a willful violation and a repeated violation.

• A willful violation exists where evidence shows that *(a)* the employer committed an intentional and knowing violation of the Act, and knows that such action constitutes a violation, or *(b)* even though the employer was not consciously violating the Act, he was aware that a hazardous condition existed and made no reasonable effort to eliminate the condition.

• A repeated violation is where a second citation is issued for a

39

violation of a given standard or the same condition which violates the general duty clause. A repeated violation differs from a failure to abate in that repeated violations exist where the employer has abated an earlier violation and, upon later inspection, is found to have violated the same standard.

The vast majority of citations that have been issued by OSHA so far have been for alleged violations of the occupational safety and health standards and have been classified as nonserious. Relatively few citations are issued for alleged violations of the general duty clause.

Citations

When an investigation or inspection reveals a condition which is alleged to be in violation of the standards or general duty clause, the employer may be issued a written citation which will describe the specific nature of the alleged violation, the standard allegedly violated, and will fix a time for abatement. Each citation, or copy thereof, must be prominently posted by the employer at or near the place where the alleged violation occurred for at least three days, or until corrected. All citations will be issued by the area director or his designee and will be sent to the employer by certified mail.

A "Citation for Serious Violation" will be prepared to cover those violations which fall into the "serious category." This type of violation *must* be assessed a monetary penalty.

A "Citation" is used for nonser-ious violations which *may* or *may not* carry a monetary penalty. A citation may be issued to the employer for employee actions which violate the safety and health standards.

A notice, in lieu of a citation, is issued for de minimis violations that have no direct relationship to safety and health. Unlike the citation, the employer is not required to post this notice.

Penalties

In proposing civil penalties for citations, a distinction is made between serious violations and all other violations. There is no requirement that a penalty be proposed when a violation is not a serious one, but a penalty *must* be proposed for a serious violation. In either case, the maximum penalty that may be proposed is $1000. In case of willful or repeated violations, a civil penalty of up to $10,000 may be proposed. Criminal penalties may be imposed on any employer who, among other things, willfully violates a standard and that violation causes death to any employee. There are no penalties for de minimis violations. See Schedule of Penalties, Table 2-A.

Penalties may be proposed for an alleged violation even though the employer immediately abates or initiates steps to abate the alleged violation. However, actions to abate should be favorably considered when determining the amount of adjustment applied for "good faith."

Nonserious violations

For nonserious violations, the

TABLE 2–A
SCHEDULE OF OSHA CITATIONS AND PENALTIES

Violation	Penalty (maximum under law)
Nonserious	$1,000 fine
Serious	$1,000 fine
Willful (no death) or repeated violations	$10,000 fine
Willful (death results)	$10,000 fine and/or six months jail
Willful (death results) Second violation	$20,000 fine and/or one year jail
Failure to correct cited violation	$1,000 per day fine
Receiving advance notice of inspection	$1,000 fine and/or six months jail
Failure to post official documents	$1,000 fine
False documents	$10,000 fine and/or six months jail
Hampering or assaulting federal inspector	$5,000 fine and/or three years jail
Assaulting inspector with a deadly weapon	$10,000 fine and/or 10 years jail
Murder of inspector	$10,000 fine and/or life in jail

penalty may range from 0 to $1000 for each violation. An "unadjusted penalty" is based on the gravity of the alleged violation. Three factors are used; all require professional judgment.

• The *severity* of injury or illness most likely to result

• The *probability* or likelihood that an injury or illness would result from the alleged violation

• The *extent* to which the standard is violated. (Here there are two factors involved—*(a)* standards pertaining to the workplace, and *(b)* standards pertaining to employee procedures.)

Penalty reductions

The "unadjusted penalty" may then be adjusted downward to 50 percent, depending on the em-

ployer's "good faith," size of business, and history of previous violations. Here is how this works. A reduction of up to 20 percent may be given for "good faith." Evidence of good faith includes awareness of the OSHAct and any overt indications of the employer's desire to comply with the Act. A reduction of up to 10 percent may be given for business size measured in terms of the number of employees. A reduction of up to 20 percent may be given for a favorable history regarding previous violations. Normally, such history is based on the employer's past experience under the OSHAct. However, in certain cases, the employer's past history under other federal or applicable state safety and health statutes may be considered. The penalty adjustment factors are ap-

41

plied to the "unadjusted penalty" to find the "adjusted penalty."

The "adjusted penalty" is further reduced by 50 percent (the abatement credit) if the employer corrects the violation within the abatement period specified in the citation. This reduction is made at the time the proposed penalty is calculated to determine the proposed penalty to be assessed for each violation.

Serious violations

The law requires that any employer who has received a citation for a serious violation must be assessed a proposed civil penalty of up to $1000 for each such violation. Due to the severity of a serious violation, the amount of the proposed penalty for each cited serious violation is usually calculated from a base of $1000 (the unadjusted penalty), which is the maximum penalty allowed. The unadjusted penalty may then be adjusted downward by up to 50 percent, depending on the employer's good faith, size of business, and history of violations, just as in the case of nonserious violations. However, the additional 50 percent "abatement credit" applicable to nonserious violations is *not* applicable to serious violations.

Imminent danger

Penalties may be proposed in cases of imminent danger even though the employer immediately eliminates the imminence of such danger or initiates steps to abate such danger. If the danger is abated, the situation can be reduced in gravity to the "serious" category, and some to the "nonserious"

42

category—all dependent on what was done to remove the imminence and how much of the hazard was removed.

Notice of proposed penalty

Once the proposed penalties for the serious and/or nonserious violations have been calculated, the "Notification of Proposed Penalty" is prepared and sent to the employer by certified mail.

Notice of failure to correct

The Act provides that any employer who fails to correct an uncontested violation within the abatement period may be assessed a proposed penalty of up to $1000 for each day that the violation continues after the expiration of the abatement period. This penalty provision can be applied when a followup inspection discloses that the employer has not abated a violation for which a citation has been issued and the citation and proposed penalty have become final. In such cases, a "Notification of Failure to Correct Violation and of Proposed Additional Penalty" is used to notify the employer; this again is sent by certified mail.

Time for payment of penalties

When a citation and/or proposed penalty is uncontested, the payment is due after the lapse of 15 working days following receipt of the penalty notice. When a citation and/or penalty are contested, the payment (if any) is not due until the final order of the Occupational Safety and Health Review Commission or of the appropriate Circuit Court of Appeals is issued.

Contesting Proposed Enforcement Actions

An employer has the right to contest an OSHA action if he feels that such action is not justified. The employer may contest a citation, a proposed penalty, a notice of failure to correct a violation, the time allotted for abatement of an alleged violation, or any combination of these. An employee or authorized employee representative may contest only the time allotted for an alleged violation.

Prior to going through the formality of initiating a contest, employers should request an informal hearing with the area director or the assistant regional director. Many times such informal sessions will resolve the questions and the issues; and the formal contested case proceedings can be avoided.

If the informal conference fails to resolve the dispute between OSHA and the employer and the latter elects to contest the case, it must be remembered that affected employees or the authorized employee representative are automatically deemed to be parties to the proceeding. In contesting an OSHA action, the employer must comply with the following that apply to the specific case:

• Notify the area office which initiated action that he is contesting. This must be done within 15 working days from receipt of OSHA's notice of proposed penalty; it must be sent by certified mail. If the employer does not contest within 15 working days after receipt of the notice of proposed penalty, the citation and proposed assessment of penalties are deemed to be a final order of the Occupational Safety and Health Review Commission and are not subject to review by any court or agency and the alleged violation must be corrected within the abatement period specified in the citation.

• If any of the employees working on the site of the alleged violation are union members, a copy of the notice of contest must be served upon their union.

• If any employees who work on the site are not represented by a union, a copy of the notice of contest must either be posted at a place, where the employees will see it or be served upon them personally.

• The notice of contest must also contain a listing of the names and addresses of those parties who have been personally served a notice and, if such notice is posted, the address(es) where the notice was posted.

• If the employees at the site of the alleged violation are not represented by a union and have not been personally served with a copy of the notice to contest, posted copies must specifically advise the unrepresented employees that they may be prohibited from asserting their status as parties to the case if they fail to properly identify themselves to the Commission or the hearing examiner prior to the commencement of the hearing or at the beginning of the hearing.

• There is no specific form for

43

the notice of contest. However, such notice should clearly identify what is being contested—the citation, the proposed penalty, the notice of failure to correct a violation, or the time allowed for abatement of the violation or any combination of these.

If the employer contests an alleged violation in good faith, and not solely for delay or variance of penalties, the abatement period does not begin until the entry of the final order by the Review Commission.

When a notice of contest is received by an area director from an employer or from an employee representative, he will file with the Review Commission the notice of contest and all contested citations, notice of proposed penalties, or notice of failure to abate.

Upon receipt of the notice of contest from the area director, the Commission will assign the case a docket number. Ultimately, a judge (hearing examiner) will be assigned to the case and will conduct a hearing at a location reasonably convenient to those concerned. OSHA presents its case and is subject to cross examination by other parties. Then the contesting party presents his case and is also subject to cross examination by other parties. Affected employees or an authorized employee representative may participate in the hearings. The decision by the judge will be based *only* on what is in the record. Therefore, if statements are unchallenged, the statements will be assumed to be fact.

Upon completion of the hearings, the judge will submit the record and his report to the Review Commission. If no commissioner orders a review of a judge's recommendation, such recommendation will stand as the Review Commission's decision. If any Commissioner orders a review of the case, the Commission itself must render a decision to affirm, modify, or vacate the penalty. The Commission's orders become final 15 days after issuance, unless stayed by a court order.

Any person adversely affected or aggrieved by an order of the Commission may obtain a review of such order in the U.S. Court of Appeals if sought within 60 days of the order's issuance.

Small Business Loans

The Act enables economic assistance for small business. It amends the Small Business Act to provide for financial assistance to small firms for changes that will be necessary to comply with the standards promulgated under the OSHAct or standards promulgated by a state under a state plan. Before approving any such financial assistance, the Small Business Administration (SBA) must first determine that the small firm is likely to suffer substantial economic injury without such assistance. Additional information is contained in "OSHA Fact Sheet for Small Business on Obtaining Compliance Loans," Bulletin No. OSHA 2005, available from local OSHA offices.

Maintain Safe and Healthful Working Conditions

CHAPTER 3

A successful occupational safety, health, and industrial hygiene program tries to prevent accidents and illnesses and the resulting injuries, property losses, and lost time. Although inadequate supervision, weak maintenance programs, and human imperfections prevent any employer or manager from being completely assured that all hazards are under control, he can have a successful accident prevention program to the degree that he follows the seven basic elements outlined in the Introduction, and the suggestions of the previous chapter.

Maintaining safe working conditions is one of those seven steps. Hazards must be identified before they can be eliminated. This chapter covers types of hazards to be controlled, the basic technique for controlling them, and how to maintain control by anticipating accidents and preventing them before they happen. The practical application of this technique is given in controlling housekeeping hazards (the second part of this chapter).

Control or Eliminate Common Hazards

Any workplace or job site can be broken down into operational groupings or categories which have common hazards associated with them. There are six such operational categories which can be applied to any and every business in every industry. They form a "system" or "plan" to use to assure that hazards are identified and can be reviewed on a periodic and regular basis by self-inspection.

Such a plan or system can be developed into an inspection survey guide or program which any employer can develop and personalize to his own operations.

For example, an employer or other designated person evaluates the hazards of the operations at the workplace or job site. He then considers additional hazard potentials as other elements of production are added to the work location. This provides a better understanding of the relationship of each category to the group of categories.

45

The following describes how these categories apply individually and collectively:

Workplace hazards

There are certain basic safety and health hazards which apply strictly to the workplace, as it exists alone, as a building or other work location. These hazards include such things as floors or other working surfaces, housekeeping, floor and wall openings, entrances and exits, sanitation, illumination, fire, ventilation, and others.

Machine and equipment hazards

When machines and equipment are added to the workplace, new hazards come into play. And, there are standards which cover the avoidance of these risks, for example those covering machine guarding, operational techniques, special safety devices, inspection and maintenance, mounting, anchoring, grounding, and other protection.

Material hazards

Materials that are utilized, processed, or applied on the job add additional new elements of hazard. There are specific standards covering materials that yield dangerous vapors, fumes, or mists, ignitable and/or explosive dusts, and other atmospheric contaminants. There are standards for safe storage and handling of compressed gases, flammable and combustible liquids, and requirements for safe storage and handling of more stable materials used in production processes. (Standards are covered a little later in this chapter.)

Worker-employee hazards

When the employee is "added" to a workplace, special precautions and requirements become important. What type of medical and first aid service is required? What type personal protective equipment and devices must be provided? Are licenses or other accreditation documents required? What about special training and educational requirements? (These factors are discussed in subsequent chapters.)

Power source hazards

The power source creates an additional source of hazards which must be controlled. Electrical, pneumatic, hydraulic, steam, explosive-actuated and other sources of power have standards applicable to their safe use and application.

Operation hazards

Some standards cover a special process or a special industry where multiple hazards combine to create problems. For example, welding, cutting and brazing, spray finishing, abrasive blasting, and dip tanks are hazardous processes.

As he becomes more familiar with this system or conceptual breakdown, a person may wish to elaborate on it and add special requirements relative to his own operation.

Standards and Administration

All operational categories can be considered at one time or taken individually. For example, machines and equipment can be considered first and other categories at other times. Some industries and establishments that do not have many hazards in some categories should concentrate first on those most applicable in their own operations. Periodic inspections should be scheduled to make sure conditions are kept under control.

Special operations should be checked to determine if specialists, such as industrial hygienists, design engineers, or chemists, should be called in for consultation.

Standards

Most operations and equipment are covered by the OSHA regulations, discussed in the previous chapter, and by standards published by private organizations, such as the following:

ANSI. American National Standards Institute, 1430 Broadway, New York, N.Y. 10018

NFPA. National Fire Protection Association, 470 Atlantic Ave., Boston, Mass. 02210

ASHRAE. American Society of Heating, Refrigeration and Air Conditioning Engineers, Inc. 345 East 47th St., New York, N.Y. 10017.

HEW. U.S. Department of Health, Education and Welfare, Public Health Service, Washington, D.C. 20203

ASME. American Society of Mechanical Engineers, Inc., United Engineering

Center, 345 East 47th St., New York, N.Y. 10017

CGA. Compressed Gas Association, Inc., 500 Fifth Ave., New York, N.Y. 10036

ACGIH. American Conference of Governmental Industrial Hygienists, 1014 Broadway, Cincinnati, Ohio 45202

If a specific operation is not covered by an existing standard, yet it could cause an accident or illness, the requirements specified to control other related or similar exposures may be used as a reference. It is not likely that every conceivable hazard will ever be covered by a specific standard or government regulation.

Administrative regulations

In addition to federal and state safety and health standards outlined in the previous chapter, there are administrative requirements which must be met. For example, every employer must display an OSHA poster stating the rights and obligations of employees and employers (Figure 2-2); keep injury, illness, and exposure records; report fatalities and multiple hospital injury cases (5 or more) within 48 hours; post an annual summary of injuries and illnesses; post notices of citation at or near the place of violation; notify employees of applications for variances; and other duties discussed in Chapter 2.

Housekeeping Hazards Can Be Controlled

In the plan for controlling or eliminating common hazards just discussed, it was suggested that the evaluation of hazard potential for any business should start from a base point. The base point recommended was the workplace or jobsite.

The concept is that any workplace or jobsite has certain inherent hazards just by reason of its existance. And, in such hazard evaluation, the workplace must be considered as if no machinery, materials, workers, power sources, or speical processes were in existence yet. This could be likened to a jigsaw puzzle made up of six parts, which when all put together constitutes the safety picture for the business.

Since Chapter 9 through 13 in this handbook cover the actual discussion of many hazards and their controls, only one basic operation—housekeeping, a potential source of workplace hazards—will be considered.

Since the status of industrial housekeeping can affect all categories of hazards in most instances, this is a good program to start with for most businesses. A good housekeeper has a system and a place for everything. A good housekeeper is organized in thinking and in action. System and organization are two of the most important elements in any safety program.

There always will be some dirt and disorder when making a product or providing a service, but if it is allowed to accumulate, sooner or later there may be production problems, higher employee turnover, and increased accident rates. People do not enjoy working in areas that are dirty, disorderly, crowded and booby-trapped with hazards. Good housekeeping is good business.

Results and cost benefits

Immediate and long-range results can be obtained by developing and instituting a well-planned and well-administered housekeeping program.

Reduced operating costs. Once an operation is clean and a housekeeping system has been established, less time and effort are required to keep it clean.

Increased production. Once obstacles to production are removed, orderly and businesslike methods can function without undue interference or delay.

Improved production control. It is easier to keep track of material and parts. Checking operations and recording data are easier.

Conservation of materials and parts. Unused materials, including spoilage and scrap, are easily and quickly removed to the proper place.

Save production time. The need to search for tools, parts, or materials is eliminated.

Better use of floor space. Clear areas permit free movement of operators and afford repairmen easy access to machines and equipment.

Open aisles permit faster traffic with fewer collisions.

Lower accident rates. Sufficient work space and elimination of tripping, slipping, bumping against, being struck by, dropping, and caught between object hazards mean fewer injuries.

Higher employee morale. Clean surroundings and comparative freedom from injury improve employee attitudes.

Reduced fire hazards. Good housekeeping is vital to effective control of fire hazards.

These ten points summarize the main benefits of instituting a good housekeeping program. What is involved in an effective program is not just a "pushbutton" effort, but also an orderly arrangement of operation, tooling, equipment, storage, and supplies.

Inspecting for housekeeping

Inspection is to industrial housekeeping as quality control is to production. Inspection does not mean spying, or trying to find something wrong or someone to blame. It is a way of determining whether or not everything is satisfactory, and of uncovering physical defects and unsafe plant conditions. Thus it is essential to preventive maintenance as well as housekeeping. Inspection may also reveal unsafe practices of employees, such as operating equipment without authority or proper training, operating equipment at unsafe speeds, using unsafe materials handling methods, not using proper personal protective equipment, or not wearing proper clothing. When these are found, the supervisor must correct the practices and later follow up to see that they stay corrected. The next chapter covers the subject of safety inspections in detail.

Techniques. Many managers check housekeeping conditions constantly as they go through their operations areas. Supervisors should tour their departments at the end of their shift. Still others delegate the job to an assistant who, in turn, reports back. If accidents or general inspection rcords show up housekeeping deficiencies, supervisors should be made aware of the problem and the need to give special attention to certain areas.

Progress chart. Regardless of the inspection method used, a housekeeping progress chart, displayed where all can see it, will keep employees informed and help maintain interest in the housekeeping program.

This chart can include such items as orderliness and cleanliness by operation or area; scrap and rubbish control and removal; in-process materials handling and storage; proper light and ventilation; and safety considerations (such as breaches of safety rules, and blocking of stretchers or fire extinguishers).

Building structure

Walls and columns should not be used as places to hang clothing, job tickets, calendars (with pinup girls), rags, or similar objects. As soon as one item is up, it invites others and soon the area is covered.

Window sills are not storage spaces. Even one milk carton left on a ledge is too many; it is the start toward making the ledge a catchall.

Shelves should be kept in repair. They are generally not

Figure 3-1.—Floor loads, approved by the building official, shall be marked on plates which shall be supplied and affixed by the owner of the building, or his agent, in a conspicuous place in each space to which they shall relate. Where the same type of material is stored regularly, walls and columns can be marked to indicate the height to which material can by piled without exceeding the allowable floor load.

strong enough to be used for storage. Even if they are, a person could be struck by a falling object.

Stairs and exits should be well lighted and designated as exits. They should be free from obstructions, have treads and handrails in good repair, and should be kept clean. Even a small bolt or pencil could cause a bad injury if dropped and not picked up immediately.

Working platforms 4 feet or more high must have strong guard rails, and also toeboards if there is a danger of material falling off.

Maintenance of buildings, structures, and equipment is an extremely important phase of a good housekeeping program. Dust and dirt accumulations on walls and ceilings, windows and skylights, and lighting fixtures reduce illumination. Faded, washed-out paint also reduces the value of identifying colors on machinery, and equipment and the distinctive colors on signs. When these conditions develop, painted surfaces should be scheduled for cleaning and repainting as necessary.

Floors and working surfaces

Floors and working surfaces should be level and kept as slip-

50

proof as possible and should be swept or vacuumed clean at regular intervals. They should be free of imbedded chips and accumulated drippings. Material that might drip or spill can be collected in drip pans or gutters, or be deflected by spash guards. If these liquids do get on the floor, nonflammable absorbent materials should be readily available. Sawdust, because of its combustibility, should never be used as an absorbent.

Areas that are not swept or vacummed as often as the operation demands should be reported. Cracks, splinters, ruts, and breaks should be repaired as soon as they are discovered. Maintenance must be proper for the type floor, working surface, or floor covering in the various operating areas.

Safe load limits for floors must be determined, marked, and maintained. See Figure 3-1.

Aisles. The type and amount of traffic that uses an aisle determines how wide it should be.

Aisles should be clearly defined by painted lines, plastic tape, guard rails, or some other satisfactory means. These also serve as a constant reminder that aisles are to be kept clear. The portion of the aisle used for pedestrian traffic should be clearly designated.

Blind corners and intersections should have (a) shatterproof mirrors placed so that traffic in an aisle is visible to anyone approaching from the side, and (b) warning signs to remind all traffic of an impending corner or intersection.

Bump rails should be installed along shop offices, storage areas, and machinery as protection from trucks. Aisles should not be used as storage space.

Exit signs and directions must be kept visible.

Storage facilities

Storage areas should be marked off and kept separate. Their location and size should be realistic and contribute to the efficiency of the department. Incoming and outgoing materials should be kept separate. Operating space should not be used for long-term storage of small parts, such as nuts, bolts, washers, and springs.

Piling, stowing, or stocking should be safe, orderly, and neat. Piles, under sprinkler heads, should be kept at a distance prescribed by fire regulations (usually 18 in. minimum; 24 to 36 in. recommended—doubled when stock is piled more than 15 ft high).

If materials handling equipment is kept in the work area, a storage space should be allotted for it also. Containers should never be overloaded and should be kept in good repair.

Cabinets used for storage should be kept closed and the permissible contents labeled on the outside. Other material should be kept out. Flammable material should be stored in a separate area or building as prescribed by fire regulations.

Racks are considered the best solution for storing ladders. Never leave a ladder on the floor or against a wall or partition.

See details on pages 172-178.

51

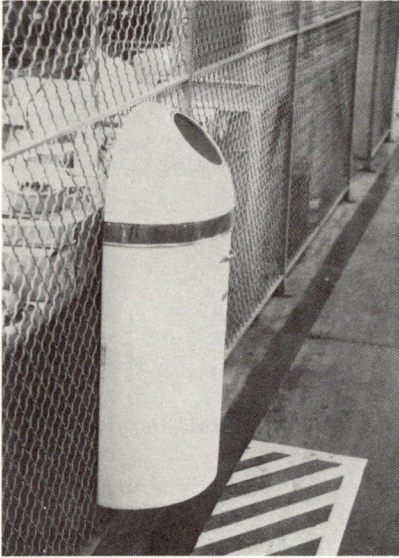

Figure 3-2.—Make it easy for employees to dispose of trash properly.

Employee facilities

Personal belongings belong in lockers, not in work areas. Lockers should be cleaned out and inspected periodically to prevent unhealthful or unsanitary accumulations. When toxic materials are involved, special precautions concerning clothing and cleaning should be initiated.

Toilet rooms should be well ventilated, well lit, and provided with separate receptacles for disposal of towels and smoking materials. Floors, toilets, and lavatories should be cleaned at least once each shift, and more often if necessary. Lavatories should have hot and cold water, soap, and towels. Sterilizers or deodorizing disinfectants should be used for offensive odors, but it is better to eliminate the source rather than cover them up with strong-smelling disinfectants.

Cleaning and sanitation material and equipment should be stored carefully to prevent waste. Such equipment should not be stored in toilet or locker rooms, in halls, in corners, or on stairways.

Eating space, with adequate bottle and trash receptacles should be provided and cleaned at least once each shift. Employers should not be allowed to litter the floor. (See Figure 3-2.) Drinking fountains and any beverage or food-dispensing facilities should be cleaned each shift.

Cuspidors should be provided in work areas for those who chew tobacco or use snuff.

Smoking must be strictly prohibited when it could be a fire hazard. NO SMOKING signs must be posted to warn employees. Both eating and smoking should also be prohibited in areas containing toxic materials that could enter the mouth by these means.

Machines and equipment

Like aisles and storage areas, machine areas should be marked off, and machines placed so that operators are not exposed to aisle traffic. Allow space for maintenance and repair, for storage of incoming and outgoing processed stock, and for tools, jigs, and fixtures needed. Tools, jigs, and fixtures are not to be left around haphazardly, but kept in racks and containers.

Work benches are not catchalls for personal belongings, fixtures, tools, spare parts, and odds and ends. They should be kept clean and in good repair. Good lighting, tool racks, cabinets for extension

52

cords and air hoses, and storage space for dies and gages and other equipment should be provided. Trash containers should be spotted conveniently throughout the work area.

Scrap, clips, cuttings, and dust should be dumped directly from machines or benches into containers and not onto the floor. Containers should be emptied regularly. Spoiled work should be removed for reclamation at least twice a shift. Finished or in-process items should be fed and taken away efficiently and safely and not left on the machine or on the floor. Both scrap and work containers should be rated for specified loads and removed promptly when full.

Proper lubrication of machinery makes housekeeping easier. Follow manufacturer's instructions for lubrication. Train oilers to avoid using too much oil or grease, and to clean up any spilled oil or grease before they leave an area.

Machines should be kept clean and in proper adjustment and repair. Machines should be painted colors that make it as easy as possible to see the work, spotlight the point of operation, and service. (More about colors later.)

Special-hazard areas should be clearly marked by signs (NO SMOKING, HARD HAT AREA, etc.) that must be kept clean and legible at all times. Obsolete signs should be removed. (Accident prevention signs are discussed later.)

Machine guarding is discussed in Chapter 9, Machines and Tools.

Fire protection

Fire protection depends on the nature of the operation and the character of the materials. Knowing this, make sure that portable fire extinguishers of the correct type and adequate size are available in sufficient quantity. Make sure that all employees know where the extinguishers are located and know how to use them. See details in Chapter 12, Fire Protection.

Yards and grounds

Yards and grounds should be carefully maintained. Follow the same principles recommended for inside storage. Keep bulk materials in neat, well-trimmed piles; keep grass and weeds cut down. Traffic ways and railway spurs should be kept free of trash and obstructions, and be kept well drained to reduce ice and mud and the chances of skids, slips, or other mishaps. (Keep abrasive or ice-melting material handy, just in case.) Traffic lanes help control vehicle and pedestrian movement.

Use of color

Industry uses color in four ways:

• For plant interiors and general equipment

• For marking hazards and identifying particular equipment

• For identifying content of piping systems

• For accident prevention signs

Marking hazards and protective equipment. The *Safety Color Code for Marking Physical Haz-*

53

ards, American National Standard Z53.1, provides for uniform marking of physical hazards and for identifying and indicating the location of fire and other protective equipment. Use of color is not a substitute for safeguards, but rather is a supplement to them. Be sure to check latest regulations for in-plant use, shipping, or consumer protection. In summary they are as follows:

RED identifies fire protection equipment, danger, and emergency stops on machines.

YELLOW is the standard color for (*a*) marking hazards that may result in accidents from slipping, falling, striking against, etc.; (*b*) flammable liquid storage cabinets; (*c*) a band on red safety cans; (*d*) materials handling equipment, such as lift trucks and gantry cranes; and (*e*) radiation hazard areas or containers (AEC still requires purple). Black stripes or "checker board" patterns can be used.

GREEN designates the location of first aid and safety equipment (other than fire fighting equipment). (Also see blue.)

BLACK AND WHITE and combinations of them in stripes or checks are used for housekeeping and traffic markings. They are also permitted as contrast colors.

ORANGE is the standard color to highlight dangerous parts of machines or energized equipment, such as exposed edges of cutting devices and the inside of (*a*) movable guards and enclosure doors, and (*b*) transmission guards.

BLUE is used on informational signs and bulletin boards not of a safety nature. (If of a safety nature, use green.) It also has railroad uses.

REDDISH-PURPLE identifies radiation hazards; check AEC regulations.

The piping in a plant may carry harmless, valuable, or dangerous contents, and therefore it is highly desirable to identify different piping systems. American National Standard A13.1, *Scheme for Identification of Piping Systems,* specifies standard colors for identifying pipelines and describes methods of applying these colors to the lines. The contents of pipelines are classified:

Classification	*Color*
Fire protection	Red
Dangerous	Yellow
Safe	Green
Protective materials (e.g., inerting gases)	Bright blue

The proper color may be applied to the entire length of the pipe or in bands 8 to 10 in. wide near valves, pumps, and at repeated intervals along the line. The name of the specific material is stenciled in black at readily visible locations such as valves and pumps.

Piping less than $3/4$ in. diameter is identified by enamel-on-metal tags.

The code also recommends highly resistant colored substances for use where acids and other chemicals may affect paints.

54

Safety Inspections

Purposes and Objectives of Inspections

Finding unsafe conditions and work practices by means of inspection and promptly correcting them is one of the best methods for management to prevent accidents and to safeguard employees. Management also demonstrates to employees its interest and sincerity in accident prevention. Likewise, failure to find and promptly correct unsafe conditions destroys worker confidence in the employer's sincerity.

Inspections also help sell the safety program to employees. Each time an inspector or an inspection committee passes through the work area, management's interest in safety is shown. Regular plant inspections encourage individual employees to inspect their immediate work area.

In addition, inspections enable safety personnel to come in contact with individual workmen and to enlist their help in eliminating accidents. Frequently, the workmen are able to point out unsafe conditions that may otherwise go unnoticed and uncorrected. When an employee's suggestions are acted upon, he realizes that he has contributed to the safety of the plant, and that his cooperation is appreciated.

Hazard categories

The following hazard categories have specific safety, industrial hygiene, and health regulations which must be met (see the previous chapter). Those that apply should be built into any inspection program.

WORKPLACE HAZARD CATEGORIES

- Walking-Working Surfaces
- Ladders
- Scaffolds
- Stairs and Stairways
- Life Safety-Exit Facilities
- Fire Suppression Equipment
- Electrical Wiring, Apparatus, and Equipment
- Boilers, Heating and Cooling Equipment, Pressure Vessels, Piping
- Illumination
- Ventilation—General and Comfort
- Sanitation
- Accident Prevention Signs
- Labels and Other Markings

55

- Color Codes—Marking Physical Hazards
- Medical Services and First Aid

MACHINES AND EQUIPMENT HAZARD CATEGORIES

- Machine Guarding Requirements—General
- Hazardous Mechanisms to be Guarded
- Safeguarding the Point of Operation
- Guarding Power Transmission Appartus
- Standard Guard Design and Construction
- Guarding Rotating and Cutting Mechanisms
- Guarding Grinding and Sanding Machines
- Guarding In-Running Nip Points
- Guarding Miscellaneous Rotating or Revolving Mechanisms
- Guarding of Reciprocating Mechanisms
- Shearing and Cutting Mechanisms
- Piercing Machinery
- Hand and Portable Tools and Other Hand-Held Equipment
- Elevators, Powered Platforms, Manlifts, Hoisting Machinery, Equipment and Devices
- Material Handling Machines and Equipment

MATERIALS HAZARD CATEGORIES

- Hazardous Materials—Atmospheric Contaminants, Carcinogens
- Radiation Hazards
- Oxidizing Agents
- Combustible Dusts
- Flammable and Combustible Liquids

- Acids, Caustics and Other Harmful Substances
- Compressed and/or Liquefied Gases
- Explosives and Blasting Agents
- Materials Storage and Handling
- Hazardous Materials—Data and Record Requirements

POWER SOURCE HAZARD CATEGORIES

- Electrical Power—Grounding, fusing, cords, locations, warnings, installations, hazardous locations
- Pneumatic Power—Hoses, relief devices, maintenance, guards, piping, containers
- Hydraulic Power—Hoses, relief devices, approved containers, locations, maintenance
- Steam Power—Relief devices, approved systems, heat protection, construction, maintenance
- Explosive Actuated Power—Tool construction, training, standard procedures

SPECIAL PROCESS HAZARD CATEGORIES

- Surface Preparation, Finishing, Preservation
- Welding, Cutting, Heating, Brazing
- Concrete Forms, Shoring
- Steel Erection
- Tunnels, Shafts, Caissons, Cofferdams, Compressed Air
- Demolition
- Blasting and Use of Explosives
- Power Transmission and Distribution
- Trenching
- Rollover Protective Structures
- Many Others by Industry Group

In addition, good inspection observes employee unsafe acts and practices, such as:

- Reaching into moving machinery
- Misuse of machinery, hoisting equipment, and lifting services
- Improperly lifting and carrying
- Horseplay
- Violation of "No Smoking" areas
- Performing unauthorized work
- Improper wearing apparel
- Nonuse or improper use of personal protective equipment and devices
- Disregard for good personal hygiene
- Improvised methods and practices
- Operating equipment at unsafe speeds and without guards
- Operating without proper training

Inspection planning and checklists

Systematic inspection is the basic tool for maintaining safe conditions and checking unsafe practices. Each company, plant, or department should develop its own checklist. A sample checklist, stressing work areas or work practices or both, is shown in Figure 4-1. Checklists or inspection survey guides based on OSHA safety and health regulations are available and can be used to conform to government requirements.*

In preparing for an inspection, it is advisable to analyze all accidents (including no-injury accidents and near misses, if possible)

so that special attention can be given those conditions and those locations known to be accident producers. Attention should be paid to the number of accidents, the departments in which they occurred, the types of accidents, the agencies involved, the nature of the injuries, and, most important, the accident causes.

A well-planned inspection depends upon knowing where to look and what to look for.

Types of Inspections

Once the check points have been determined and a systematic procedure worked out, there are four ways that the work area can be inspected. Safety inspections can be classified as follows:

- Periodic inspections
- Intermittent inspections
- Continuous inspections
- Special inspections

Periodic inspections

Periodic inspections are those scheduled to be made at regular intervals. It is advisable to schedule inspections for the entire plant, for certain operations, or for certain types of equipment. Such inspections may be made

---

*The National Safety Council has *OSHA Standards Inspection Checklists* for 29 CFR 1910, general industry, and 29 CFR 1926, construction. Also *OSHA Inspection Survey Guide and Survey Report* is published by Management Research and Development Institute, 321 E. William Street, Wichita, Kansas 67202.

SAFETY INSPECTION CHECKLIST

Plant or Department_____ Date_____

This list is intended only as a reminder. Look for other unsafe acts and conditions, and then report them so that corrective action can be taken. Note particularly whether unsafe acts or conditions that have caused accidents have been corrected. Note also whether potential accident causes, marked "X" on previous inspection, have been corrected.

(√) indicates *Satisfactory*   (X) indicates *Unsatisfactory*

1. FIRE PROTECTION
Extinguishing equipment................ ☐
Standpipes, hoses, sprinkler heads
and valves .... ☐
Exits, stairs and signs.................. ☐
Storage of flammable material .... ☐

2. HOUSEKEEPING
Aisles, stairs and floors................ ☐
Storage and piling of material .... ☐
Wash and locker rooms .... ☐
Light and ventilation................... ☐
Disposal of waste .... ☐
Yards and parking lots................. ☐

3. TOOLS
Power tools, wiring .... ☐
Hand tools........................... ☐
Use and storage of tools .... ☐

4. PERSONAL PROTECTIVE EQUIPMENT
Goggles or face shields .... ☐
Safety shoes .... ☐
Gloves .... ☐
Respirators or gas masks .... ☐
Protective clothing .... ☐

5. MATERIAL HANDLING EQUIPMENT
Power trucks, hand trucks .... ☐
Elevators .... ☐
Cranes and hoists .... ☐
Conveyors .... ☐
Cables, ropes, chains, slings .... ☐

6. BULLETIN BOARDS
Neat and attractive .... ☐
Display changed regularly .... ☐
Well illuminated...................... ☐

7. MACHINERY
Point of operation guards............. ☐
Belts, pulleys, gears, shafts, etc........ ☐
Oiling, cleaning and adjusting .... ☐
Maintenance and oil leakage........... ☐

8. PRESSURE EQUIPMENT
Steam equipment .... ☐
Air receivers and compressors .... ☐
Gas cylinders and hose................ ☐

9. UNSAFE PRACTICES
Excessive speed of vehicles .... ☐
Improper lifting .... ☐
Smoking in danger areas .... ☐
Horseplay .... ☐
Running in aisles or on stairs.......... ☐
Improper use of air hoses............. ☐
Removing machine or other guards...... ☐
Work on unguarded moving machinery... ☐

10. FIRST AID
First aid kits and rooms................ ☐
Stretchers and fire blankets .... ☐
Emergency showers .... ☐
All injuries reported .... ☐

11. MISCELLANEOUS
Acids and caustics .... ☐
New processes, chemicals and solvents..... ☐
Dusts, vapors, or fumes .... ☐
Ladders and scaffolds................. ☐

Signed_____

USE REVERSE SIDE FOR DETAILED COMMENTS OR RECOMMENDATIONS

REP. 500PADS16301   PRINTED. U.S.A.   STOCK No. 129.95

Figure 4-1.—Detailed inspection checklist helps inspector find unsafe acts and hazards before they can trigger an accident.

monthly, semiannually, annually, or at other suitable intervals. Some types of equipment, such as elevators, boilers, unfired pressure vessels, and fire extinguishing equipment are required by law to be inspected at regular intervals.

All periodic inspections should be well planned so they can be made systematically with dispatch and efficiency.

General inspections. In many companies or plants, it is an established policy to make a general inspection of the entire premises each year, except for those departments or equipment that are scheduled for more frequent inspections.

A general inspection should cover those places that "no one ever visits" and "where no one ever gets hurt." Many of these out-of-the-way places (where lives have been lost) are located overhead where it is difficult to see a hazard from the shop floor.

Inspections should not be confined to those places where serious injuries have occurred. This is why no-injury accidents and near-accidents should be checked—they will often point to causes of future injuries.

General inspections are particularly valuable before reopening a plant after a long shutdown or prior to a safety campaign or contest. However, they should not take the place of other types of inspections. A general inspection should not signal that all responsibility has ended for the continual inspection throughout the year, nor should known hazards be ne-

Figure 4-2.—Detection of unsafe conditions and unsafe acts should not depend only on formalized inspections. Any observed hazard should be corrected.

glected until such general inspections are made (Figure 4-2).

Inspection of buildings and physical plant should be conducted at least yearly. Questions to ask are:

• Do the physical conditions noted contain potential accident hazards?

• What practical measures can and should be taken to remove such hazards or failure points?

• Is there any catastrophe hazard present that warrants immediate corrective measures?

An adequate checklist will call

59

attention to the following items and conditions:

Gounds—Parking lots, roadways, and sidewalks need frequent inspection for cracks, holes, breaks and tripping hazards.

Loading and shipping platforms—Loading and shipping platforms and docks get severe use from trucks and heavy traffic. Guard rails are needed to prevent damage from tail ends of trucks.

Outside structures—Small isolated buildings should be inspected the same way as the large plant facilities. Plant fencing should be inspected for damage by trucks, cars, and other causes such as corrosion.

Railroad tracks—Railroad sidings that are company-owned require regular inspection for washouts, condition of rails, clearance, missing joint plates and spikes, rotted ties, and the need for additional ballast.

Floors—Floors, regardless of construction, should be carefully inspected especially in areas subject to heavy traffic. Slippery floors should receive special study. Here are check points.

- ✔ Are the surfaces wearing too rapidly?
- ✔ Is shrinkage present?
- ✔ Is the surface damaged?
- ✔ Are there slippery surfaces?
- ✔ Are there hazards to trucking or walking, such as holes or unguarded openings?
- ✔ Are there indications of cracks, sagging, or warping?

Courtesy Macy's, Jamaica.

Figure 4-3.—Permanent aisles and passageways should be appropriately marked. Stock must be kept out of aisles and fire lanes.

- ✔ Are replacements necessary because of deterioration?

Stairways—Stairways should be checked to determine that:

- ✔ Are treads and risers in good condition and of uniform width and height?
- ✔ Are standard handrails provided and are they in good condition and secure?
- ✔ Is lighting on stairs satisfactory?
- ✔ Is there any material stored on stairs?

Housekeeping—The general housekeeping throughout the plant should be checked to see if it is satisfactory. Aisles should be marked off with painted lines and material kept out of aisle width (Figure 4-3).

60

Electrical installations—The National Electrical Code, published as Std. No. 70 by the National Fire Protection Association, is the principal standard.

Wiring—The physical plant survey should include a periodic inspection of all plant wiring.

Transformers - Switchboards—The survey should include an examination of transformer and switchboard installations.

Elevators—The efficient and safe operation of plant elevators is an important physical survey determination.

Roofs—A specialized item in the survey of the physical plant is roof anchorage.

Chimney stacks—A thorough annual inspection of chimneys and stacks by a competent person will uncover minor damages before they progress too far and become critical.

Overloading of floors—Experience has shown that the most serious floor-loading problems encountered arise from the tendency of management to look upon warehouse or storage space as nonproductive.

Fire inspection—One of the greatest hazards to an industrial plant is fire. Consequently, a rigid system should be set up for periodic inspection of all types of fire protection equipment. Such inspections should include water tanks, sprinkler systems, standpipes, hose, fire plugs, extinguishers, and all other equipment used for fire protection. A systematic schedule of inspections should be closely followed and an accurate record kept of each piece of equipment inspected and tested. See Chapter 12, Fire Protection.

Along with this scheduled inspection, a careful survey should be made of new equipment needed. Recommendations should be made for replacement of defective and obsolete equipment, as well as the purchase of any additional equipment. As new processes and products are added to the manufacturing system, new fire hazards may be introduced that require individual treatment and possibly special extinguishing devices. Be sure to follow through.

Surveys should also include all means of egress from the building. All exits, stairs, fire towers, fire excapes, halls, fire alarm systems, emergency lighting systems, and places that are seldom used should be thoroughly inspected to determine their adequacy and readiness for emergency use.

Elevators and pressure vessels—Another type of periodic inspection is that required by state and local laws which include elevators, boilers, and unfired pressure vessels. Such equipment, however, is not usually inspected by the company employees, but rather by the insurance carrier or other specialist. A prearranged schedule of inspections as required by law should be followed.

The proper persons should be notified well in advance of such inspections so arrangements can be made to shutdown the equipment until the inspection is completed. It is wise to have supplementary inspections made more frequently by a qualified company employee.

Chains, ropes and slings— Chains, wire and fiber ropes, and other equipment subject to severe strain in handling heavy equipment and materials should be inspected at regular intervals. A careful record should be kept of each inspection. Some state and OSHA regulations require such inspection and records. This type of equipment should be stenciled or marked to show the date and result of the latest inspection.

Tools and appliances—Many types of equipment and processes require periodic inspections if they are to be operated safely and efficiently. Some companies require that all tools and appliances be returned to a central storeroom after use each day, where they are carefully checked and repaired before being reused. A thorough "knockdown" inspection is made every few weeks.

Other periodic inspections— Pay attention to conditions causing falls, the second largest cause of accidental death. Fall-causing hazards include: slippery, wet, oily, and worn floors; ice and snow on walks and platforms; stumbling hazards; loose material underfoot; worn or broken treads on stairs; insecure scaffolds and platforms; stairs, scaffolds, and platforms with no handrails; defective ladders or ladders not suited to the job; open elevator shaftways; unguarded floor openings and manholes; and inadequate lighting.

Other types of equipment such as cranes, hoists, presses, ladders, and power trucks require periodic inspection. Equipment used in the field, such as in construction or excavation, also requires frequent and periodic inspections. Such inspections should be determined by the proper plant executives and the safety director should prepare a working schedule so that the correct intervals of inspection can be maintained. Operators of all powered materials handling equipment should be required to make a daily inspection of their equipment.

It is desirable to have periodic inspections by some organized group or an individual in every department. These should be made at one-month intervals. The report of these inspections should be made by departments for the plant manager or company president. In some plants, foreman are required to make such inspections and reports.

Intermittent inspections

Inspections made at irregular intervals are the most common. They may include an unannounced inspection of a particular department, piece of equipment, or small work area. Such inspections tend to keep the supervisory staff alert to find and correct unsafe conditions before being found by a safety inspector or the government compliance officer.

The need for intermittent inspections is frequently indicated by accident tabulations and analysis. If the analysis shows an unusual number of accidents for a particular department or location, or an increase in certain types of injuries, inspections should be made to determine the reasons for the increase and what is necessary to make corrections.

Intermittent inspections may be made not only by a safety department, but also by supervisors, safety committees, and individual workmen who know how to inspect.

Continuous inspections

Many companies have set up a system of continuous inspection so that selected employees observe certain equipment and operations. A maintenance man may spend some time each day roaming about the plant or department, continuously observing operations and making adjustments and minor repairs. These men are also responsible for the safe operation of assigned machines and help train employees in safe practices. This method of inspection makes these men familiar with the exact condition of each part of the machines covered. It enables them to point out weaknesses and defects long before they become serious or hazardous.

A continuous system of inspection of both the condition and the proper use of personal protective equipment is especially desirable. A constant check on goggles, respirators, safety shoes, gloves and other protective clothing, as well as any other personal protective devices, will assure proper use and maintenance of this equipment.

It should be obvious that any machine or equipment operator should continuously be on the alert for any defects and immediately report things needing attention.

Systematic inspection by management

Supervisors should continuously make sure that tools, machines, and other equipment are maintained properly and are safe to use. They should check the proper use and maintenance of machine guards and safety devices. To do this effectively, they use systematic inspection procedures and may delegate authority to other responsible persons.

Toolroom personnel should inspect all hand tools to see that they are kept in safe condition. Some companies require portable electric tools to be turned in to the electrical department for a periodic check, as discussed earlier.

Inspection programs should be set up for new or redesigned equipment, material, and processes. Nothing should be put into regular operation until it has first been checked for hazards, its operation studied, additional safeguards installed (if necessary), and safety instructions developed.

Serious injury has occurred because this procedure was not followed. This is also a good time to train employees in safe operation. It takes less time and effort now than if done later.

Crew leaders are often given the responsibility for inspecting equipment and for seeing that their men observe safe practices. The supervisor should make certain that these inspections are up to required standards.

By systematic planning and organization, the supervisor can carry out his primary function—supervision. He spot checks periodically to make sure assignments are being carried out, that safety precautions are being observed, and that equipment is running efficiently and safely.

Special inspections

Special inspections are sometimes necessary because of the installation of new buildings or remodeling of old ones, or because of new processes or equipment. Also, special inspections may be made during special campaigns, such as fire prevention week or waste elimination campaigns.

Accident investigation requires special inspections by the investigation committee and the safety professional. These should be made with the same thoroughness and determination to control accident causes, as are periodic inspections.

Health surveys. Wherever there is a real or suspected health hazard, a special inspection should be made to determine the extent of the hazard and the precautions necessary to provide and maintain safe conditions. These inspections usually require air sampling for the presence of toxic fumes, vapors or gases, and dust; testing of materials for toxic properties; or the testing of ventilation and exhaust systems for efficiency of operation. See photograph on page 3.

Also, periodic physical exami-

nations should be made of employees exposed to occupational disease hazards. This is a form of special inspection conducted by the medical staff as a disease-control measure.

Testing for toxic substances and physical examinations require special equipment not always available to the inspector or doctor. In such cases, assistance can frequently be secured from consultants, the industrial hygiene division of the state department of labor or health, from industrial hygienists employed by insurance companies, or from individual consultants.

In connection with the health survey, industrial sanitation facilities should also be inspected with special attention given to personal service facilities. This is particularly important where there is possible exposure to certain toxic substances, such as lead, carcinogens, and where personal cleanliness and effective decontamination procedures is a large factor in controlling occupational disease.

OSHA requirements for health hazard control are becoming increasingly more strict. Therefore, all potential health hazards must be identified and kept under constant surveillance to assure no excessive exposure of personnel. (See Chapter 13, Industrial Hygiene.)

Overhead inspections. Special inspections for overhead hazards are very important in the control of accident causes. Hazardous conditions often exist because of loose objects that may fall from

Figure 4-4.—Here is a bad situation that should have never happened in the first place: the exhaust of diesel-powered air compressor was placed right in front of the fresh air intake for a building. Employees and supervisors must be constantly on the alert.

building structures, cranes, roofs, and other overhead locations.

During such inspections, it is not uncommon to find loose tools, bolts, pipe lines, shafting, pieces of lumber, windows, electrical fixtures, and other objects that are potential accident-makers. Overhead inspections frequently disclose the need for repairs to skylights, windows, cranes, roofs, and other installations that affect the safety not only of employees,

but the operation of the machines and equipment.

Also, the existence of overhead devices that require frequent attention, adjustment, cleaning, oiling and repairing demands that employees occasionally visit such locations. Inspections are necessary to determine that all reasonable safeguards are provided and safe practices observed.

Every overhead job, including inspections at elevated locations, should be carefully planned so it can be done safely.

Accident investigations. It is obvious that every accident and near-accident that occurs should be thoroughly investigated as soon as possible to find its actual and contributing causes in order to prevent a recurrence. During the investigation, a special inspection of the scene of accident is essential. The inspection should be made only by persons who have at least a basic knowledge of accident prevention. A photograph of the scene will be helpful for future reference. A full discussion of accident investigation is in Chapter 6.

Inspections initiated by employee complaints to OSHA.

Under the Occupational Safety and Health Act, employees have the right to request the government to inspect any employer's facilities if they believe a violation of the OSHA safety and health regulations exists which may be dangerous to their health. This complaint should be filed in writing; however, it may be made by telephone or in person.

When such complaints are made, the names of the employees will be kept anonymous if they so wish. And, employers will usually not be informed of the complaint until the government inspector contacts them on the work premises. During such inspections, employee representatives also have the right to accompany the inspector to see that the items complained of are observed. This was detailed in Chapter 2, OSHA Considerations.

Inspection Is Everyone's Business

Inspectors should know how to locate hazards and have the authority to act when they discover out-of-line conditions or procedures.

Inspectors must set an example. It is of utmost importance that the inspector be properly equipped with personal protective equipment, protective clothing, and any necessary apparatus as required by the nature of the inspection. It is difficult for an inspector to persuade the chipper or grinder to wear goggles when he himself is not wearing them, or to "sell" safety shoes to workmen exposed to foot injuries, if he does not wear them, or to have workmen use respirators unless he wears one when in dusty and noxious atmospheres. It is essential to "practice what you preach."

Many people are involved in safety inspections. Their duties are outlined in this section.

Company or plant management

Safety inspection should be part of the duties of all company or plant supervisors, including top management or the superintendent. Interest in the safety program is naturally stimulated if the top manager stops to make a casual inspection as he passes through various departments. Should an unguarded hazard or an unsafe practice be discovered, he should immediately confer with the supervisor to see that the condition is corrected.

First-line supervision

The first-line supervisor or

foreman is one of the most important "inspectors" in any organization, because the foreman spends practically all of his time on the job. He is in constant contact with his employees and thoroughly familiar with all the hazards that could develop in his department. He should be on the alert at all times to discover and correct unsafe conditions and practices. (See Figure 4-4.)

Because the foreman's duties require him to give close supervision over his entire department, most of his inspections should be of the continuous type. However, he should make general inspections at regular intervals to make sure that all hazards have been properly safeguarded and that the safeguards are being used. Also, he should keep a close check on the periodic inspection of elevators, pressure vessels, and other special equipment.

The foreman's responsibility requires that he continuously observe surrounding conditions and the tools used by the men in his crew. He must realize that conditions change continually and that unsafe conditions can arise at any time.

Engineering and maintenance.

The mechanical engineer and the person responsible for maintenance should make frequent trips through the plant. Necessary work orders for guards or for correcting faulty equipment can be written up on the spot. They should watch for unsafe conditions and unsafe practices and report such conditions to the foreman for correction. If, unfortunately, these two officials are not

safety minded, their attitude will have a detrimental effect on all safety activities.

Employees

Employees who are constantly on the alert are of great value in preventing accidents. In some companies, the rank-and-file employee is encouraged to inspect his workplace each day and to report any hazardous conditions to his supervisor. Employees who are safety conscious will always look for conditions that may cause an injury to themselves or to others. (See Figure 4-4.)

Each operator may be required to inspect his workplace and the equipment or machinery he uses. He does this each day and immediately reports defects that he is not authorized to correct. In departments having a number of similar machines, a specialist should make regular inspections and perform the necessary adjustments and repairs.

When maintenance employees are working in various departments and observe hazards which should be corrected, they can be of great help if they report such hazards to the foreman of the department. Management should convince foreman that maintenance men can be of great help in locating and correcting hazardous conditions. Similarly, foremen should encourage mechanics to come to them with their suggestions.

Scheduled inspections are necessary in plant storehouses, warehouses, and toolrooms. Here new machinery, tools, appliances, and materials enter the plant and are checked before being requisitioned. The storekeeper and toolroom men can, therefore, contribute a great deal to safety by inspecting and checking for simple defects.

Safety committees

Smaller companies or plants without a full-time professional can have regular inspections made by a safety committee. No one person with only part-time safety responsibilities can cover an entire operation without help.

The group which makes inspections should give equal consideration to accidents, fire, and health exposures. If the committee is familiar with the history of accidents in the company or plant, the approach to the inspection can be planned all the more intelligently. The group should, therefore, have knowledge of all accidents which have occurred in the area involved before it makes an inspection.

It is common practice to have accidents investigated by the safety committee or special accident investigation committee. Where a company or plant is sufficiently large to justify such procedure, a standing committee for the investigation of accidents will not only get more details concerning the accidents, but will also become more safety-minded. Such a committee should make a complete inspection of the department as well as of the scene of the accident.

Paper work for the inspection committee should be kept to a minimum. Checklists prove useful for helping the inspection committee to cover all details, but since checklists cannot be all-inclusive, they should not be relied upon entirely in general inspections. The committee can also develop report forms to simplify the problem of summarizing findings at the end of a tour.

The committee will become less effective if meetings are allowed to degenerate into "gripe sessions." The committee chairman should be strong enough to keep the committee productive.

The inspection committee should not attempt to deal with technical problems best analyzed by trained observers or special testing methods. It should, however, be able to recognize the need for expert assistance where the possibility of danger exists, as in exposure to dusts, gases, vapors, or fumes, which may create either health or fire and explosion hazards, and which call for scientific testing methods to measure the hazards and determine remedies.

Safety professionals

When companies are large enough to have a full-time safety professional, he must spearhead inspection activity. He is in his most productive role during safety or health inspections. Industrial hygienists and fire protection specialists also have a great stake in inspections.

Companies or plants that are too small to employ a full-time safety director must depend a great deal on inspections made by maintenance men and supervisors, committees, or a part-time safety director. Many plants depend entirely on inspection service supplied by casualty insurance inspectors and state factory inspectors. More frequent inspections are usually necessary, however, than can be provided by these agencies.

When using inspectors, it should be recognized that a person qualified for general types of inspection work may not be qualified to make inspections of a special type. For instance, pressure equipment (such as boilers, autoclaves, digestors, air receivers, etc.) and materials handling equipment (such as cranes, hoists, elevators, chains, and slings) are among the most hazardous equipment used in industry. The inspection and testing of such equipment requires engineering knowledge and training. It should be determined that the inspector understands properly the hidden physical hazards involved and is competent to pass judgment on such equipment.

When chemicals are used, it is important that they measure up to exact standards of purity and packaging as otherwise they may cause fires or other bad accidents. In such cases, the chemist and safety inspector should work in close cooperation.

Also, in industries where toxic and corrosive substances are present (such as dust, gases, vapors, and liquids), the safety engineer or inspector should be specially trained to make sure he is familiar with the properties of these substances and with methods of controlling the hazards involved. An industrial hygienist should be consulted.

Inspection Procedures

Inspection procedures vary considerably in different companies and plants and with the types of inspections made.

Inspectors, of course, should know thoroughly all their company's safety and health rules and policies. They should also be familiar with approved federal and state safety codes, laws, and regulations, and municipal ordinances that apply to assure the safety and health of workers. The fire protection requirements that are applicable to the particular plant should also be mastered. Usually federal and state codes, laws, and regulations set up minimum requirements only. It is frequently necessary to exceed these requirements to comply with company policy and secure maximum safety.

Before making periodic or special inspections, it is highly important to make a complete analysis of company and department accident statistics and accident causes discussed in Chapter 5.

Aids for the inspector should include:

- Inspection checklists
- Inspection report forms
- Job safety analysis cards

Starting the inspection

Three basic steps of how to inspect are: (*a*) contact the department head or job superintendent and solicit his help, (*b*) observe all conditions for compliance with established standards (use checklists), and (*c*) observe all operations for any unsafe acts or violations of safety rules.

69

Step 1. Before an inspection is started in any operation, the inspector should first contact the person in charge. Because the department head is responsible for the safety of his workers and whatever unsafe conditions that may exist, his participation and cooperation in making an inspection can expedite any needed corrections.

It is good practice, therefore, for the supervisor to accompany the inspector while covering his operations.

Step 2. Inspections should be systematic and thorough. Use a checklist, as suggested earlier in this chapter. All locations that may contain hazards should be inspected. This could be accomplished by following the operations from the receiving of raw material until the finished product is shipped or stored. Occasionally, it is advisable to vary the route followed. In one plant, where inspections follow the flow of material, the route is reversed for inspections.

Step 3. Making observations and taking notes.

Notes made at the time unsafe conditions and practices are discovered form the basis of a complete report to be prepared later. The inspector should not depend on his memory and write notes after he has left the particular department or returned to his office.

It is important to secure all the information needed to describe the hazard found. The exact location and other pertinent data necessary to make a complete report,

such as suggestions for correcting the condition, should also be given.

It should not be the aim of the inspector to pick up numerous trivial items merely to make his report look good. A good report concisely covers all details that must be corrected. The "who, what, where, when, and why" technique has proven effective. A long list should not ordinarily result in an area that is regularly inspected. If safety work has been neglected, however, a long list can be expected.

Each recommendation should be discussed with the supervisor so he will be fully informed. In this manner, an agreement can usually be reached as to the relative importance of recommendations.

When the supervisor fully understands what is required, he may be able to make corrections in a very short time. Sometimes, the supervisor's promise to correct the hazard is not carried out, so it is advisable to include all items in the report.

The same procedure should be followed when accompanying a government inspector on an OSHA inspection. See Chapter 2.

Making the report

Every inspection must be followed by a clearly written report. Inspection reports are usually of three types:

1. The emergency report—made without delay whenever immediate corrective action is necessary.

2. The routine report—covers

70

those observations of unsatisfactory (nonemergency) conditions that should be corrected.
This report should be made no
later than one day after the
inspection, and should also include the emergency report.
3. The periodic report—lists all
items of previous routine reports for a given period. This
report is actually a summary of
safety activity and accomplish-
ments.

Handling recommendations

Recommendations should be
listed in the order in which they
were discovered, or grouped by
the operation according to the individual responsible for their
compliance. Where possible, a
definite time limit for compliance
should be set for each recommendation, or they should be grouped
according to their importance and
marked "urgent," "important," or
"desirable."

Recommendations approved by
the management should become a
part of the plant improvement or
maintenance program, be assigned job numbers, and pass
through the regular channels for
such matters. Supervisors should
report progress in complying with
the recommendations to management (or the general safety committee) at regular intervals. Also,
the inspector should make periodic checks of all recommendations until they have been completed. In this manner, he may be
able to make suggestions for the
best method of eliminating the
hazard and, at the same time, prepare a progress chart for his own
information.

Condemning equipment

A practical system can be established for taking certain materials
and equipment from service because of unsafe conditions. Special danger tags (see Figure 4-5)
can be employed effectively in
preventing the use of equipment
or materials that have become unsafe through wear, abuse, or defects. Only the inspector who
places the tag should be permitted to remove it and only when he
is satisfied that the hazardous
condition has been corrected.

No equipment or materials
should be placed out of service
without notifying the person in
authority in the department affected. A shutdown, to avoid what
might appear to be a possible hazard, might interrupt work at great
expense without actually afford-
ing any protection. Consequently,
authority to condemn equipment
should be exercised with a great
deal of caution by the inspector.

Making night inspections

When a firm or plant is working
on a two- or three-shift schedule,
it is desirable occasionally to have
inspectors working with each
shift. Safety conditions vary considerably after dark, because of
artificial illumination. The inspectors should make sure that
adequate illumination is provided
and that the lighting system is
maintained in a satisfactory condition.

When regular night shifts are
not employed, night working conditions frequently go unobserved.
An inspector should be assigned
to make occasional plant visits to
check on the safety of mainte-

nance men, firemen, and others who may be required to work after dark. Also, an occasional trip with the night watchman to observe the conditions under which he works should be made.

Making photo inspections

A combined safety inspection and supervisor training session can be the result of a quick tour through an operation with an instant-developing camera, a camera equipped to take 2x2-inch color slides, or even a lightweight video tape recorder.

Even smaller companies or contractors can personalize safety meetings and develop a safety training resource and capability in this way. An instant-developing camera easily records unsafe conditions or practices, and the pictures then become a subject for the supervisor's meeting.

Checking plans and specifications

As discussed in Chapter 1, plans and specifications for major modifications or alterations and for new buildings and equipment should be checked to discover and correct conditions that may otherwise be built into the building and its equipment and that would later result in modification costs, and in injuries or penalties under OSHA regulations. Instructions and warnings on products sold by the company should be reviewed both for safe manufacture and safe use of product.

The employer should also make sure that company policies and applicable standards are followed in purchase specifications for new

materials and equipment or for modification to existing equipment.

For instance, when a new chemical is requested, the purchaser should see that information on flammability, toxicity, and similar properties is obtained from the manufacturer. This handbook outlines requirements and performance standards that should be used to evaluate the safety specifications for purchases of industrial equipment and raw materials.

A safety program for a contractor who is working on company premises can be tied into the contract documents.

Safety observation plan for work practices

A safety inspection of work practices requires the active participation of all supervisory personnel. In a safety observation plan, supervisors (particularly the first-line supervisors directly responsible for employee safety) observe work while it is in progress. These observations cover the use of tools, materials, and equipment, as well as any unsafe method or order of procedure in performing an act which indicates a lack of planning or failure to take into account all the circumstances surrounding the particular job. In fact, all employees, even experienced workers, should be checked once in a while.

One reason why safety observation has not been more fully utilized is that safety personnel and supervisors seldom develop or improve their power of observation. Another is that planned safe-

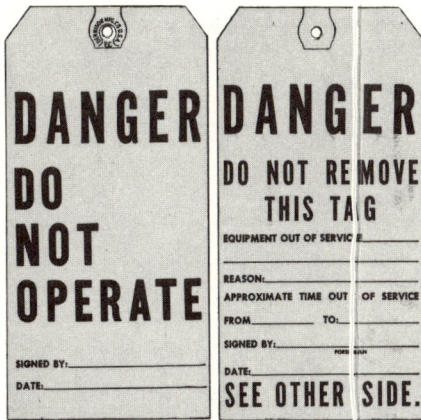

Figure 4-5.—Front and back views of typical tag used when equipment is taken out of service because it has become unsafe.

ty observation involves a little more effort than the incidental safety observation.

There are several factors necessary for effective safety observation. The inspector must:

1. *Be selective.* An inspector might inspect a department first, for safety; second, for improvement of operations; third, for training needs; and so on.

2. *Know what to look for.* The more a person knows about a job and a worker's responsibilities, the better an observer he will be.

3. *Practice observing.* The more often a person looks with the conscious intention to observe, the more he will see at each fresh trial. Like all skills, observation improves with practice.

4. *Keep an open mind.* One way

to increase open-mindedness is not to judge facts in advance. The inspector must not deny the fact, no matter what conclusion it may seem to lead to. The inspector must keep his mind open, at least until he has all the facts.

5. *Do not be satisfied with general impressions.* A clean shop, or a careful routine, may still contain hidden hazards.

6. *Guard against habit and familiarity.* Asking the questions What, Where, When, How, and (especially) Why will often help uncover the real meaning of the situation.

7. *Record observations systematically.* All notes should be dated, with space for comment on action taken and on results of the action. The notebook can serve both as a reminder and as a record of progress.

When a person is observed performing an unsafe practice or act, the supervisor is faced with a problem in human relations skills. The worker should be left with the feeling that he has been corrected in a decent way. Here are a few pointers that might help:

1. Make your contact private.

2. Be firm, but friendly.

3. Determine his reasons for acting unsafely.

4. Explain what and why.

5. Review the safe alternative.

6. Get agreement on future practice.

73

Accident Records and Reports

CHAPTER 5

Legal Requirements

Occupational injury and illness records are now required in nearly every business establishment. Under the Occupational Safety and Health Act, a majority of employers are required to maintain certain records of work-related injuries and illnesses. In addition to these records, many employers must make reports to state compensation boards or their insurance companies. In contest and award programs, reports based upon the American National Standard Z16.1 may be needed.

Small employers are faced with two tasks—maintaining these required records and maintaining records necessary for an effective safety program. Often, a good recordkeeping system needs more data than that contained in the required forms. This chapter deals with both aspects of recordkeeping.

OSHA and state requirements

Specific legal requirements are beyond the scope of this publication because they differ from industry to industry, from state to state, and with time. The appropriate federal and state authorities should be contacted to obtain the most current requirements.

OSHA requirements. An outline of the general recordkeeping requirements of the Occupational Safety and Health Administration, as they exist at the time of publication, is presented here. The requirements and definitions are subject to change, and the individual states, which may implement the Act in their own jurisdiction, may make modifications. To determine the most current definitions and requirements, contact the appropriate state or federal authorities.

Regulations issued under the Occupational Safety and Health Act of 1970 require all establishments subject to the Act to maintain records of recordable occupational injuries and illnesses, occurring on or after July 1, 1971. Such records consist of:

- A log of occupational injuries and illnesses, OSHA Form 100

• A supplementary record of each occupational injury or illness, OSHA Form 101

• An annual summary of occupational injuries and illnesses, OSHA Form 102

• Records of employee exposure to toxic substances or harmful physical agents as required by the standards and regulations. (See the section on Special Records and Reports, near the end of this chapter.)

• Other records such as logs of inspections and tests as required by the various safety and health standards

OSHA Forms 100, 101, and 102 are available at all Bureau of Labor Statistics regional offices. If your state has an OSHA-approved plan, be sure to check for any special recordkeeping requirements.

In an effort to relieve small businesses from the recordkeeping requirements, OSHA has ruled that employers who had no more than seven employees during the calendar year immediately preceding the current calendar year need not maintain the log, the supplementary record, nor prepare or post the summary. It should be noted however, if an employer has more than one unit and a total of more than seven employees, records must be kept for all individual units. However, if an employer, regardless of size, has been notified in writing by the Bureau of Labor Statistics that he has been selected to participate in the statistical survey of occupational injuries and illnesses, then he will be required to maintain the log and make reports for the period of time specified in the notice.

Beginning January 1, 1975, employers are required to record job-related injuries and illnesses on revised log form, OSHA No. 100. The revised OSHA Form 102 summary will not be used until January 1976 when the injury and illness data for 1975 is summarized. These are shown in Figure 5-1.

OSHA Form 100—Log of occupational injuries and illnesses. Each employer is required to maintain in each establishment a log of all recordable occupational injuries and illnesses for that establishment (see Figures 5-1 and 5-2). However, the log may be maintained at a place other than the establishment or by means of data processing equipment provided that:

• There is available at the place where the log is maintained sufficient information to complete the log to a date within six working days after receiving information that a recordable case has occurred.

• At each of the employer's establishments there is available a copy of the log of that establishment complete and current within 45 calendar days.

Every employer is required to enter each recordable occupational injury and illness on the log as early as practicable, but no later than six working days after receiving information that the re-
(Text continues on page 78.)

EXAMPLES OF COMPLETED RECORDKEEPING FORMS

OSHA NO. 100

LOG OF OCCUPATION...

Form Approved
OMB No. 44R 1453

RECORDABLE CASES: You are required to record information about: every occupational death; every nonfatal occupational illness; and those nonfatal occupational injuries which involve one or more of the following: loss of consciousness, restriction of work or motion, transfer to another job, or medical treatment (other than first aid).
More complete definition appear on the other side of this form.

CASE OR FILE NUMBER	DATE OF INJURY OR ONSET OF ILLNESS	EMPLOYEE'S NAME (First name or initial, middle initial, last name)	DESCRIPTION OF INJURY OR ILLNESS — Nature of Injury or Illness and Part(s) of Body Affected (Typical entries for this column might be: Amputation of 1st joint right forefinger; Strain of lower back; Contact dermatitis on both hands; Electrocution—body.)	Injury or Illness Code See codes at bottom of page.	DEATHS (Enter date of death.)	LOST WORKDAY CASES — Enter a check if case involved lost workdays.	Enter number of days AWAY FROM WORK due to injury or illness.	Enter number of RESTRICTED WORK ACTIVITY due to injury or illness.	NONFATAL CASES WITHOUT LOST WORKDAYS (Enter a check if no entry was made in columns 8 or 9 but the case is recordable, as defined above.)	TERMINATIONS OR PERMANENT TRANSFERS (Enter a check if the entry in columns 9 or 10 represented a termination or permanent transfer.)
(1)	(2) Mo./day/yr.	(3)	(6)	(7)	(8) Mo./day/yr.	(9)	(9A)	(9B)	(10)	(11)
1	2-7-75	George O. White	Amputation - First joint, Right forefinger	10		✓				
2	4-23-75	Robert T. Burns	Fracture - Right arm	10		✓	13	10		✓
3	5-8-75	William W. William	Sunstroke - Body	25		✓	10	1		
4	5-14-75	Edward P. Smith	Second degree burn - Back of right hand	10		✓		3		
5	7-17-75	Randolph L. Davis	Rash - Both hands	21					✓	
6	9-18-75	Joseph P. Brown	Contusion - Right thumb	10					✓	

EXTENT OF AND OUTCOME OF CASES

From Recordkeeping Requirements, Rev. 1975, published by Occupational Safety and Health Administration.

76

The sample summary at left was prepared using the cases from the log above as follows:

The first line (code 10) of the summary is used to record injuries. These are identified on the log by a code 10 in column 7.

Col. 1. Total Cases: Four cases involved injuries (identified by a code 10 in column 7 of the log).

Col. 2. Deaths: There were no deaths (no entries in column 8 of the log).

Col. 3. Total Lost Workday Cases: Three of the four injury cases had a check in column 9 of the log.

Col. 4. Cases Involving Days Away From Work: Two of the four injury cases had an entry in column 9A of the log.

Col. 5. Days Away From Work: The two injury cases which had entries in column 9A of the log had a total of 23 days away from work.

Col. 6. Days of Restricted Work Activity: The two injury cases which had entries in column 9B of the log involved a total of 13 days of restricted work activity.

Col. 7. Nonfatal Cases Without Lost Workdays: One injury case had a check in column 10 of the log.

Col. 8. Terminations or Permanent Transfers: One injury case had a check in column 11 of the log.

CHECK: The sum of columns 2, 3, and 7 should equal the number entered in column 1. It does.

The same procedure is used to summarize each occupational illness category. There are entries on the lines for codes 21 and 25 because these codes were used in column 7 of the log to record the two illness cases.

On the line for code 30, Total - Occupational Illnesses, the sum of the entries for codes 21 through 29 were entered.

The totals for all occupational injuries and illnesses were entered on the line for code 31. The entries for codes 10 and 30 were added to arrive at these totals.

OSHA No. 102 | Complete no later than one month after close of calendar year. See back of this form for posting requirements and instructions. | Form Approved OMB No. 44R 1453

SUMMARY OF OCCUPATIONAL INJURIES AND ILLNESSES FOR CALENDAR YEAR 1975

Establishment:
NAME   MAIN PLANT
ADDRESS   ANYTOWN, U.S.A.

INJURY AND ILLNESS CATEGORY	CODE	TOTAL CASES (1)	DEATHS (2)	LOST WORKDAY CASES — Total Lost Workday Cases (3)	LOST WORKDAY CASES — Cases Involving Days Away From Work (4)	LOST WORKDAY CASES — Days Away From Work (5)	LOST WORKDAY CASES — Days of Restricted Work Activity (6)	NONFATAL CASES WITHOUT LOST WORKDAYS (7)	TERMINATIONS OR PERMANENT TRANSFERS (8)
		Number of entries in Col. 7 of the log.	Number of entries in Col. 8 of the log.	Number of checks in Col. 9 of the log.	Number of entries in Col. 9A of the log.	Sum of entries in Col. 9A of the log.	Sum of Col. 9B of the log.	Number of checks in Col. 10 of the log.	Number of checks in Col. 11 of the log.
OCCUPATIONAL INJURIES	10	4	0	3	2	23	13	1	1
Occupational Skin Diseases or Disorders	21	1	0	0	0	0	0	1	0
Dust Diseases of the Lungs	22								
Respiratory Conditions Due to Toxic Agents	23								
Poisoning (Systemic Effects of Toxic Materials)	24								
Disorders Due to Physical Agents	25	1	0	1	0	0	1	0	0
Disorders Associated With Repeated Trauma	26								
All Other Occupational Illnesses	29								
TOTAL—OCCUPATIONAL ILLNESSES (Sum of codes 21 through code 29)	30	2	0	1	1	0	1	1	0
TOTAL—OCCUPATIONAL INJURIES AND ILLNESSES (Sum of code 10 and code 30)	31	6	0	4	2	23	14	2	1

I certify that this Summary of Occupational Injuries and Illnesses is true and complete, to the best of my knowledge.

Signature _Jane K. Smith_
Title _Office Manager_
Date _1-9-76_

This is NOT a report form. Keep it in the establishment for 5 years.

Figure 5-1.

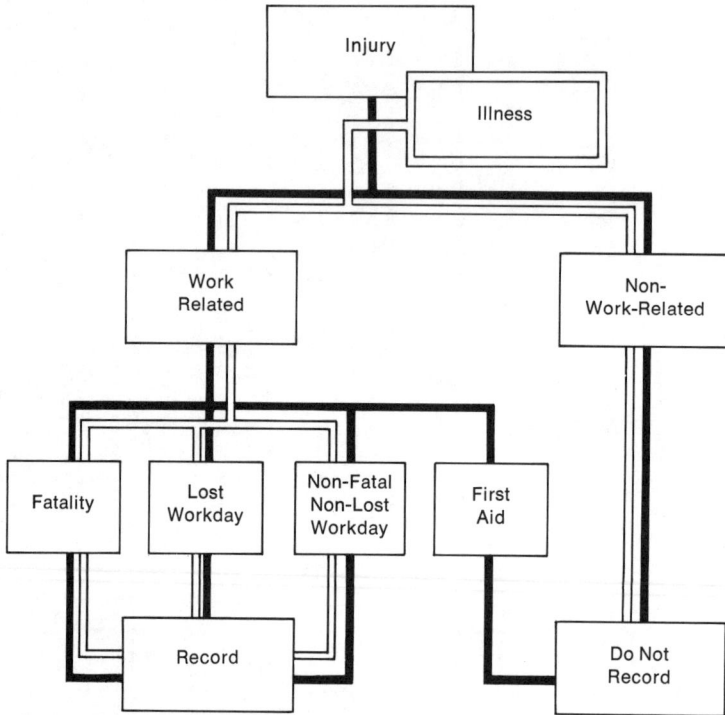

Figure 5-2.—Conditions necessary for recording occupational injuries and illnesses under the occupational safety and health act.

cordable case has occurred. OSHA Form 100 or any private equivalent may be used. The log is to be established on a calendar year basis.

OSHA Form 101—Supplementary record. Each employer is required to have available for inspection at each establishment a supplementary record of each occupational injury or illness for that establishment. This form is to be completed within six working days after receiving information that the recordable case has occurred. Other report forms are acceptable if they contain at least the information required by

OSHA Form 101. (NSC's Supervisor's Accident Report Form, shown in Figure 5-4, is an example.)

OSHA Form 102—Annual summary. Each employer is required to compile an annual summary of occupational injuries and illnesses for each establishment. The annual summary is to be based on information contained in the log for the given establishment. OSHA Form 102 must be used for this purpose (no substitutes or alternatives are allowed), and it must be completed by February 1 of each year. The summary must be certified—the em-

78

ployer or officer or employee who supervises the preparation of the summary must put his signature in the lower right-hand corner of the summary. The employer is required to post a copy of the establishment's summary in each establishment no later than February 1 and it is to remain in place for 30 consecutive calendar days after posting. See Figure 5-1 for examples of a set of completed forms.

Retention of and access to records. All three record forms must be maintained in each establishment for five years following the end of the year to which they relate.*

Records for personnel who do not primarily report or work at a single establishment, such as traveling salesmen, technicians, engineers, and the like, are to be maintained at the location from which they are paid or the base from which they operate to carry out their activities.

Employers of employees engaged in physically dispersed operations such as occur in construction, installation, repair or service activities who do not report to any fixed establishment on a regular basis, but are subject to common supervision, may satisfy the recordkeeping requirements by *(a)* maintaining records of each operation subject to common supervision (field superintendent or supervisor) in an established central place; *(b)* having the address and telephone number of the central place available at each worksite; and *(c)* having personnel available at the central place dur-

ing normal business hours to provide information from the records maintained there by telephone or by mail.

The records are to be available for inspection and copying by compliance officers, by any representative from the Bureau of Labor Statistics, by any representative of NIOSH or by any representative of a state agency accorded jurisdiction.

Reporting fatal or catastrophic events. One of the reporting requirements is that within 48 hours of an event which results in a fatality to one or more employees or results in the hospitalization of five or more employees, the employer must report the event, either orally or in writing, to the nearest OSHA office. The report should contain an explanation of the circumstances pertinent to the accident, the number of fatalities, and the extent of any injuries.

The purpose of the notice is to enable a compliance officer to be dispatched to the scene if an accident investigation is deemed necessary. The purpose of the accident investigation is to determine if there is a violation of any standard which may have contributed to the accident.

OSHA Form 103—Occupational injury and illness surveys. The Bureau of Labor Statistics is

---

*The employer should not confuse these with retention requirements for medical records of employees exposed to dangerous health substances. See Table 5-B, page 96.

charged with the responsibility for the collection, compilation, and analysis of occupational injury and illness data. To discharge this responsibility BLS conducts statistical surveys, collecting information from employers on a selective basis. Those employers selected to participate receive a written notice. The survey, which can change from year to year, utilizes OSHA Form 103, Occupational Injuries and Illnesses Survey.

OSHA Form 103, supplied by BLS, is distributed to a designated state agency and the state agency forwards the form to employers selected for inclusion in the survey. These employers must complete the form within three weeks of receipt of the form and return it to the designated state agency. Failure to comply with the reporting requirements may result in the issuance of citations and proposed penalties. The state agency will assemble the data and report designated information to the BLS headquarters in Washington, D.C. Questions concerning the survey should be directed to the state agency indicated on the survey form.

Occupational injuries and illnesses—defined

Occupational injury is any injury, such as a cut, fracture, sprain or amputation, which results from a work accident or from exposure in the work environment.

Occupational illness of an employee is any abnormal condition or disorder, other than one resulting from an occupational injury, caused by exposure to environ-

mental factors associated with his employment. It includes acute and chronic illnesses or diseases which may be caused by inhalation, absorption, ingestion, or direct contact. For purposes of information, examples of each category are given. These are typical examples, however, and are not to be considered to be the complete listing of the types of illnesses and disorders that are to be counted under each category.

● *Occupational Skin Diseases or Disorders.* Examples: contact dermatitis, eczema, rash (caused by primary irritants and sensitizers or poisonous plants), oil acne, chrome ulcers, and chemical burns or inflammations.

● *Dust Diseases of the Lungs (pneumoconioses).* Examples: silicosis, asbestosis, coal worker's pneumoconiosis, byssinosis, and other pneumoconioses.

● *Respiratory Conditions Due to Toxic Agents.* Examples: pneumonitis, pharyngitis, rhinitis or acute congestion (due to chemicals, dusts, gases, or fumes), and farmer's lung.

● *Poisoning (systemic effects of toxic materials).* Examples: poisoning by lead, mercury, cadmium, arsenic, or other metals; poisoning by carbon monoxide, hydrogen sulfide or other gases; poisoning by benzol, carbon tetrachloride, or other organic solvents; poisoning by insecticide sprays such as parathion or lead arsenate; and poisoning by other chemicals such as formaldehyde, plastics, and resins.

● *Disorders Due to Physical*

80

Agents (other than toxic materials). Examples: heatstroke, sunstroke, heat exhaustion, and other effects of environmental heat; freezing, frostbite and effects of exposure ·to low temperatures; caisson disease; effects of ionizing radiation (isotopes, X rays, radium); and effects of nonionizing radiation (welding flash, ultraviolet rays, microwaves, sunburn).

• *Disorders Due to Repeated Trauma.* Examples: noise induced hearing loss, synovitis, tenosynovitis, and bursitis, Raynaud's phenomena, and other conditions due to repeated motion, vibration, or pressure changes.

• *All Other Occupational Illness.* Examples: Anthrax, brucellosis, infectious hepatitis, malignant and benign tumors, and food poisoning.

Recordable injuries and illnesses.

Recordable occupational injuries and illnesses are those cases that result in:

• *Fatalities*—regardless of the time between the injury and death or the length of illness;

• *Lost Workday Cases*—cases, other than fatalities, that result in lost workdays;

• *Nonfatal Cases Without Lost Workdays*—cases of occupatronal injury or illness which did not involve fatalities or lost workdays but did result in *(a)* transfer to another job or termination of employment, *(b)* medical treatment, other than first aid, *(c)* diagnosis of occupational illness, *(d)* loss of consciousness, or *(e)* restriction of work or motion.

Reporting terms defined.

• *Lost workdays—days away from work* are the number of workdays (consecutive or not) on which the employee would have worked but could not because of occupational injury or illness. The number of lost workdays should not include the day of the injury or onset of illness or any days on which the employee would not have worked even though able to work.

• *Lost workdays—days of restricted activity* are the number of workdays (consecutive or not) on which, because of an occupational illness or injury:

The employee was assigned to another job on a temporary basis, or

The employee worked at a permanent job less than full time, or

The employee worked at a permanently assigned job but could not perform all duties normally connected with it.

• *Medical treatment* includes treatment administered by a physician or by registered professional personnel under the standing orders of a physician. Medical treatment *does not* include first aid treatment.

• *First aid* is a one-time treatment and subsequent observation of minor scratches, cuts, burns, splinters, and so forth, which do not ordinarily require medical care even though provided by a physician or registered professional personnel.

• *Establishment* is a single physical location where business is conducted or where services or industrial operations are performed (for example: a factory, mill, store, hotel, restaurant, movie theater, farm, ranch, bank, sales office, warehouse, or central administrative office). Where distinctly separate activities are performed at a single location (such as contract construction activities operated from the same physical location as a lumber yard), each activity shall be treated as a separate establishment.

For firms engaged in activities which may be physically dispersed, such as agriculture, construction, transportation, communications, and electric, gases, and sanitary services, records may be maintained at a place to which employees rerort each day.

Records for personnel who do not primarily report or work at a single establishment, such as traveling salesmen, technicians, and engineers, shall be maintained at the location from which they are paid or the base from which personnel operate to carry out their activities.

• *Work environment* comprises the physical location, equipment, materials processed or used, and the kinds of operations performed by an employee in the performance of his work, whether on or off the employer's premises.

• *Incidence rate* is the number of recordable injuries and illnesses per 200,000 total hours worked by all employees during period covered. The 200,000 is equivalent to 100 full-time workers at 40 hours per week for 50 weeks.

Incidence rate =

$$\frac{\text{Number of injuries and illness}}{\text{Total hours worked}} \times 200,000$$

Safety Programming Records and Reports

How to use records and reports profitably

Records of accidents and injuries are essential to efficient and successful safety programs, just as records of production, costs, sales, and profits and losses are essential to efficient and successful operation of a business. Records supply the information necessary to transform haphazard, costly, ineffective safety work into a planned safety program that controls both the conditions and the acts that contribute to accidents. Good recordkeeping is the foundation of a scientific approach to occupational safety.

A good recordkeeping system can help:

• Provide the means for objective evaluation of the magnitude of accident problems, and provide measurement of overall progress and the effectiveness of the safety program.

• Identify high accident rate units, plants, or departments and problem areas so that extra effort can be made in those areas.

• Provide data for analysis of accidents and illnesses that point to specific causes which can then be controlled by specific countermeasures.

• Create interest in safety among supervisors by furnishing information about the accident experi-

First Aid Report

Name S. D. Smith   Department Shipping

Male ☒ Female ☐ Occupation Packer   Foreman Miller

Date of
Occurrence 2-12 Time 10 a.m. Date of
                    p.m. First Treatment 2-12 Time 10 a.m.
                                                      p.m.

Nature of
Occurrence Splinter in index finger of left hand

Sent: Back to Work ☒   Doctor ☐   Home ☐   Hospital ☐

Estimated Disability 0 days

Employee's Description of Occurrence Handling wooden crates
without gloves, ran splinter into finger

Signed *M. Miller*
                                    First Aid

Issued by National Safety Council, Inc., 425 N. Michigan, Chicago, Ill. 60611
Form IS-6   Printed in U.S.A.   STOCK No. 129.26

Figure 5-3.—A First Aid Report (4 × 6 in.) is prepared at the time person comes for treatment. A report should be prepared for each case, whether minor or serious. The report serves as a record and permits quick tabulation of such data as department, occupation, and the key facts of the accident.

ences of their own departments.

• Provide supervisors and safety committees with hard facts about their safety problems so that their efforts can be concentrated.

• Measure the effectiveness of individual countermeasures, and determine if specific programs are doing the job they were designed to do.

To be effective, preventive measures must be based on complete and unbiased knowledge of the causes of accidents. The primary purpose of an accident report is to obtain such information and not to fix blame. Since the completeness and accuracy of the accident record system depend upon the information in the individual accident reports, the forms and their purpose must be under-stood by those who fill them out. Any necessary training or instructions should be available to these personnel.

Types of reports

First aid and supervisor's accident reports. The collection of injury data begins with the first aid report or the supervisor's accident report. The first aid attendant or others should know enough about accident analysis and investigation to be able to record the principal facts about each case. The physician engaged or authorized by the employer to treat injured employees should also be informed of the basic rules for classifying cases since, at times, his opinion of the seriousness of an injury may be needed for accurate recording of the case. (See Figure 5-3.)

It is recommended that the supervisor make a detailed report about each accident, even when only a minor injury or no injury is the result. For the purposes of ANSI Z16.1 or OSHA summaries, only those reports that meet the minimum severity level can be separated and tallied. Minor injuries occur in greater numbers than serious injuries, and records of these injuries can be helpful in pinpointing problem areas. By working to solve these problems, serious injuries can sometimes be prevented. Furthermore, complications may arise from the less serious injuries and the result may be serious.

The supervisor's accident report should be completed as soon as possible after an accident occurs. Information concerning unsafe acts and unsafe conditions is important in the prevention of future accidents, but often of even greater importance is information which shows *why* the injured person acted unsafely. This type of information is particularly difficult to get unless it is obtained promptly after the accident occurs.

A supervisor's report form is presented in Figure 5-4. It fulfills all of the information requirements of the present OSHA 101 form. Provision is also made for easy entry of ANSI Z16.1 information, which allows both types of recordkeeping to be included in one form without any duplication of effort. The form also includes questions in addition to those contained in the present OSHA 101 form. These questions ask for additional basic data that

should be known about each accident. It requires management participation.

Employee injury card. After cases are closed, the first aid report and the supervisor's report are usually filed by agency of injury (type of machine, tool, material, etc.), type of accident, or other factor that will facilitate use of the reports for accident prevention. Another form, therefore, is needed to record the injury experience of individual employees, such as in Figure 5-5.

This form helps supervisors remember the accident experience of individual employees. Particularly in large plants where supervisors may have many people working for them, they probably cannot recall from memory the total number of injuries—especially if the injuries are minor—which are suffered by individual employees.

The employee injury card, therefore, fills a definite need. It has space for recording such factors about the injury as date, classification, days charged, costs, and OSHA lost workdays.

Because of the importance of the personal factor in accidents, much may be learned about accident causes from studying employee injury records. If certain employees or job classifications have frequent injuries, a study of employee working habits, physical and mental abilities, training, job assignments, working environment, and their instructions and supervision may reveal more than a study of accident locations, agencies, or other factors.

SUPERVISOR'S ACCIDENT REPORT   Incident No. _____

(To be completed immediately after accident, even when there is no injury)

Company name and address _____

Plant or location address _____
(if different from above)

1. Name and address of injured _____ SSN_____ 2. Age_____
   (or ill) person

   _____ 3. Sex_____

4. Years of service_____ 5. Time on present job_____ 6. Title/occupation_____

7. Department_____ 8. Date of accident_____ 9. Time_____

10. Accident category (check)   ☐ Motor Vehicle;   ☐ Property Damage;   ☐ Fire;   ☐ Other_____

11. Severity of injury or illness   ☐ Non-disabling;   ☐ Disabling;   ☐ Medical Treatment;   ☐ Fatality

12. Amount of damage $_____ 13. Location_____

14. Estimated number of days away from job _____

15. Nature of injury or illness? _____

16. Part of body affected? _____

17. Degree of disability? _____
                          (Temporary total; permanent partial; permanent total)

18. Causative agent most directly related to accident? (Object, substance, material, machinery, equipment, conditions)

   _____

   _____

   _____

   Was weather a factor? _____

19. Unsafe mechanical/physical/environmental condition at time of accident? (Be specific)

   _____

   _____

   _____

20. Unsafe act by injured and/or others contributing to the accident. (Be specific, *must* be answered)

   _____

   _____

   _____

21. Personal factors (improper attitude, lack of knowledge or skill, slow reaction, fatigue)

   _____

   _____

   _____

(over)

Figure 5-4a.—Supervisors Accident Report, shown on this and the following page, provides a record of contributing circumstances so specific remedial action can be taken.

85

22. Personal protective equipment required? (Protective glasses, safety shoes, safety hat, safety belt)_____

Was injured using required equipment?_____

23. What can be done to prevent a recurrence of this type of accident?
(Modification of machine; mechanical guards; correct environment; training)

_____

_____

24. Detailed narrative description (How did accident occur; why; objects, equipment, tools used, circumstance, assigned duties.

Be specific)_____

_____

_____

_____

(Use additional sheets, as required)

25. Witnesses to accident_____

_____

Date prepared_____   Signature of Foreman / Supervisor_____

Department_____

SUPERINTENDENT'S APPRAISAL AND RECOMMENDATION

a. In your opinion what action on the part of injured (or ill) person or others contributed to this accident?

_____

_____

b. Your recommendation_____

_____

Date_____   Signature of Superintendent_____

FOR SAFETY OFFICE USE ONLY

Temporary Total ☐   Permanent Partial ☐   Death or Permanent Total ☐

Started losing time_____   Part of Body_____

Returned to work_____   Per cent loss or
                                                   loss of use_____

Time charge_____   Time charge_____   Time charge: 6,000 days

Compensation $_____ Medical $_____ Other $_____ Total $_____

Name and address   Name and address
of hospital_____   of physician_____

Copyright ©1969, 1972.

Issued by NATIONAL SAFETY COUNCIL • 425 North Michigan Avenue • Chicago, Illinois 60611.

A NONGOVERNMENTAL, PRIVATELY SUPPORTED, PUBLIC SERVICE ORGANIZATION   This form meets information requirements of OSHA form 101.

Form IS-IA-25M-87301   Printed in U.S.A.   Stock No. 129.21

Figure 5-4b.—Central portion is filled in by higher-level management. Bottom portion is filled in by person with safety responsibility; it contains data for computing rates and costs.

INJURY AND ILLNESS RECORD OF EMPLOYEE

S. D. Smith   845

_____(Name)_____   _____(Employee Number)_____

Occupation___Packer___ Department ___Shipping___ Date Employed _1-23-69_

Case Number	Injury or Illness	Date of Occurrence	Z16 Type (Fatal, Permanent, Temporary, Non-disabling)	Z16 Days Charged	Comp. and Other Costs	OSHA Type (Fatal, Lost Workday, Non-Lost Workday)	OSHA Lost Workdays
164	Inj.	2-12-70	Non-disab.	0	0		
192	Ill.	8-3-71	Temporary	8	0	Lost Workday	6
349	Inj.	3-16-73	Temporary	3	0	Lost Workday	1

(Reverse side may be used for remarks)

Issued by NATIONAL SAFETY COUNCIL, 425 N. Michigan Ave., Chicago, Ill. 60611

IS3 Rev.   Printed in U.S.A.   Stock No. 129.23

Figure 5-5.—An employee's injury and illness card (4 × 6 in.) is useful for seeing the injury/illness work record of an employee and for seeing any particular problems that the employee might have.

Monthly and annual summaries. The forms that have been described are prepared when the accidents occur—they are used to record the accidents and preserve information about contributing circumstances. Periodically, this information should be summarized and related to department or job exposures, so that safety work can be evaluated and the principal accident causes brought into proper focus.

A monthly summary of injury and illness cases allows for tabulating monthly and cumulative totals and the computation of ANSI Z16.1 frequency and severity rates, as well as OSHA incidence rates. Space can be provided for yearly totals and rates. This form is filled out on the basis of the individual report forms that were processed during the month. If the classification of an injury is in doubt at any time, an estimate of the outcome should be made by the company physician and the report included with completed cases in the monthly summary.

Whereas monthly summaries of injuries are prepared primarily to show the trend of safety performance during the year, annual reports are prepared so that comparisons for the longer period may be made with the experience of similar companies and of the industry as a whole.

Especially in smaller companies, monthly injury rates often show wide variations which make

87

it difficult to evaluate the safety performance correctly. If a small company has only two or three injuries in a year, the rates in the months in which these injuries occur will jump to extreme highs but, in the other months, the rates will be zero. These variations will be smoothed out in annual totals, however, and the rate for the longer period will have more significance.

Use of reports

Reports to management. Management must be increasingly interested in the accident experience of the company. Therefore, monthly and other periodic reports that show the results of the safety program should be furnished to the responsible executive. Such reports need not contain details or technical language, and they may be supplemented by simple charts or graphs to show the recent accident experience, in relation to that of the preceding period and that of companies engaged in similar work.

Bulletins to supervisors. A supervisor is primarily interested in his own department and workmen. One of the most effective ways to create and maintain the interest of supervisors and foremen in accident prevention is to keep them informed about the accident records of their departments. Injury rates, over an extended period of time, reflect the effectiveness of the supervisor in safety activities. The rates also provide the supervisors with the type of information which will help them make further reductions in injuries.

Internal publicity. Posting a variety of materials on bulletin boards is one of the best ways to maintain the interest of employees in safety. Accident records furnish many items, such as the following:

No-injury records

Unusual accidents

Frequent causes of accidents

Charts showing reductions in accidents

Simple tables comparing departmental records

Standings in contests

The agenda for employee safety meetings should particularly include information about the outstanding injury and illness problems, frequent unsafe practices, hazardous types of equipment, and similar data disclosed by analysis of the accidents that have occurred.

Reports to National Safety Council. Each year, the National Safety Council requests an annual summary report of occupational injury and illness experience from each member. These reports are tabulated to determine accident rates by industry.

An annual pamphlet, "Work Injury Rates," containing these frequency and severity rates by industry, is published by NSC and distributed to members, so that each company may compare its experience with the average experience of other companies in the same industry.

These annual summary reports are also used by the National Safety Council to evaluate its members' experience for the presentation of awards under the Council's Award Plan for Recognizing Good Industrial Safety Records.

Specific recordkeeping systems

During the period that industry is becoming accustomed to the OSHA recordkeeping system, there is no link to historical records, and therefore, measures of progress or lack of it will not be available. The National Safety Council, at this time, supports the maintenance of American National Standard Z16.1 records by establishments, at least until OSHA recordkeeping stabilizes. Another reason for using both systems is that many people feel that ANSI Z16.1 should be retained because OSHA recordkeeping does not provide an adequate measure of the severity of accidents.

Using the forms presented in this chapter, ANSI Z16.1 records can easily be kept along with OSHA records. By using a simple device, such as placing a red check mark or asterisk alongside OSHA Log entries to indicate the ANSI Z16.1 cases, little effort will be necessary to go to the individual reports and obtain the data for making annual summaries of ANSI Z16.1 cases.

ANSI Recordkeeping

Comparability of injury experience

Safety performance is relative. Only when a company compares its injury experience with (a) that of similar companies, with (b) that of the industry of which it is a part, or with (c) its own previous experience can it obtain a meaningful evaluation of its safety ac-

complishments.

To make such comparisons, a method of measurement is needed which will adjust for the effects of certain variables which contribute to differences in injury experience.

A standard method for keeping records, which provides for these variables, is given in ANSI Z16.1. First, this method uses injury frequency and severity rates which relate injuries, and days charged as a result of the injuries, to the number of manhours worked. These rates automatically adjust for differences in exposure to injury. Second, this method specifies the kinds of injuries which should be included in the rates.

Standard formulas for rates

The injury frequency rate and the injury severity rate are based on standard formulas as set forth in ANSI Z16.1

The injury frequency rate is the number of disabling injuries per 1,000,000 employee-hours worked, computed according to the following formula:

$$\frac{\text{Number of disabling injuries}}{\text{Employee-hours of exposure}} \times 1,000,000$$

The injury severity rate is the number of days charged per 1,000,000 employee-hours worked, computed according to the following formula:

$$\frac{\text{Days charged} \times 1,000,000}{\text{Employee-hours of exposure}}$$

Average days charged per disabling injury. The frequency and severity rates show, respectively, the rate at which injuries occur and the rate at which time is charged. A third measure included in the standard method shows the average severity of the injuries and may be calculated by either of the following formulas:

$$\text{I.} \quad \frac{\text{Total days charged}}{\text{Total disability injuries}}$$

or

$$\text{II.} \quad \frac{\text{Severity rate}}{\text{Frequency rate}}$$

Disabling injuries definitions

Work injuries. The standard method specifies that a work injury is any injury, including occupational disease and work-connected disability, which arises out of and in the course of employment. The following descriptions paraphrase the ANSI standard. For full details, refer to ANSI Z16.1.

Occupational disease. A disease caused by exposure to environmental factors associated with employment. Work-connected disability includes such ailments as silicosis, tenosynovitis, bursitis, and loss of hearing. Even though they are not traumatic, such disabilities, if they are work-connected, are considered work injuries.

Disabling injuries. To standardize the types of injuries to be included in the rates, certain minimum requirements have had to be established. Generally, only those injuries are included that result in death or permanent impairment, or which are sufficiently serious that the injured employee is unable to work within 24 hours of the end of the shift on which he was injured. These injuries are referred to as disabling injuries and are of four classes, as follows:

• *Death* is any fatality resulting from a work injury, regardless of the time intervening between injury and death.

• *Permanent total disability* is any injury, other than death, which permanently and totally incapacitates an employee from following any gainful occupation, or which results in the loss of use or the complete loss of any of the following in one accident: (*a*) both eyes, (*b*) one eye and one hand, arm, foot, or leg, or (*c*) any two of the following not on the same limb: hand, arm, foot, or leg.

• *Permanent partial disability* is any injury other than death or permanent total disability, which results in the complete loss or loss of use of any member or part of a member of the body, or any permanent impairment of functions of the body or part thereof, regardless of any pre-existing disability of the injured member or impaired body function.

• *Temporary total disability* is any injury which does not result in death or permanent impairment, but which results in one or more days of disability.

Note that a day of disability is any day on which an employee is unable, because of injury, to perform effectively throughout a full shift the essential functions of a regularly established job which is open and available to him. (Days include Sundays, days off, plant shutdowns, and other nonworkdays, subsequent to the day of injury. Disability days for an injury are not counted when scheduled charge(s) apply.)

Calculating employee-hours

The number of employee-hours used in calculating injury rates is the total number of hours worked by all employees, including those of operating, production, maintenance, transportation, clerical, administrative, sales, and other departments.

Employee-hours should be calculated from the payroll, or from time clock records if this method cannot be used, or they may be estimated by multiplying the total number of employee-days worked for the period covered by the number of hours worked per day.

Special cases

In addition to the usual injuries and occupational diseases, some special cases are specifically identified as work injuries, under special circumstances: inguinal hernia, back injury, aggravation of a preexisting or minor injury, skin irritations and infections, animal and insect bites, exposure to temperature extremes, and muscular disabilities if they arise out of employment duties. See NSC Industrial Data Sheet 527 or *Accident Prevention Manual*, or American National Standard Z16.1.

Days charged

Losses from work injuries are evaluated in terms of days of disability or inability to produce, either actual or potential. These losses are referred to simply as "days charged." For the first three classes of injuries—death, per-

manent total disability, and permanent partial disability—the number of days charged is a predetermined total and usually exceeds the actual time lost so as to reflect potential future losses of productive capacity. The predetermined totals are referred to as "scheduled charges." (See Table 5-A.)

This procedure is based on the philosophy of economic loss which reasons, for example, that a man who has a hand amputated will produce less during his remaining working years than a man who completely recovers from a hand injury, even though both injuries resulted in the same number of actual days lost at the time of injury. If both injuries resulted, for example, in 60 days lost at the time of the injury, the injured person who completely recovers would be charged only the 60 days, while the amputation would be charged 3000 days, the scheduled charge for this kind of injury.

Deaths and permanent total disabilities. For deaths and permanent total disabilities a scheduled charge of 6000 days is made in each case. There are no variations in this amount. If the injury is fatal, or if it results in any of the losses specified as constituting permanent total disability, the charge is the same—6000 days.

Permanent partial disabilities. For permanent partial disabilities the scheduled charges vary depending on the specific loss. For example, amputation of the index finger at the first joint has a sched-

91

TABLE 5-A
TABLE OF SCHEDULED CHARGES

A. LOSS of MEMBER—Traumatic or Surgical*

Fingers, thumb, and hand†

Amputation involving all or part of bone*	Thumb	Index	Middle	Ring	Little
			Fingers		
Distal phalange	300	100	75	60	50
Middle phalange	. . .	200	150	120	100
Proximal phalange	600	400	300	240	200
Metacarpal	900	600	500	450	400
Hand at wrist	3000				

Toe, foot, and ankle

Amputation involving all or part of bone*	Great Toe	Each of Other Toes
Distal phalange	150	35
Middle phalange	. . .	75
Proximal Phalange	300	150
Metatarsal	600	350
Foot at ankle	2400	

Arm

Any point above** elbow, including shoulder joint	4500
Any point above wrist and at or below elbow	3600

Leg

Any point above** knee	4500
Any point above ankle and at or below knee	3000

B. IMPAIRMENT of FUNCTION*

One eye (loss of sight), whether or not there is sight in the other eye	1800
Both eyes (loss of sight), in one accident	6000
One ear (complete industrial loss of hearing), whether or not there is hearing in the other ear	600
Both ears (complete industrial loss of hearing), in one accident	3000
Unrepaired hernia	50
(for repaired hernia, use actual days)	

*For loss of use of member (without amputation), use percentage of the scheduled charge corresponding to the loss of use, as determined by the physician authorized to treat the case. If the bone is not involved, use actual days lost, and classify as temporary total disability.

**The term 'above" when applied to the arm means toward the shoulder and when applied to the leg means toward the hip.

uled charge of 100 days; at the second joint, 200 days; and at the third joint, 400 days.

Scheduled charges for permanent partial injuries are shown on the accompanying Table 5-A.

For permanent partial injuries which result in loss of use of an injured member, a percentage of the scheduled charge is used which corresponds to the percentage of loss of use, as determined by the physician authorized to treat the injury.

For fatal, permanent total, and permanent partial disabilities, only the scheduled charges are used—the actual days of disability are disregarded. In some permanent partial injuries, there may be losses which exceed the scheduled charge. In either of these cases though, the number of days lost is disregarded and only the scheduled charge is used.

Temporary total disabilities. For temporary total disabilities the number of days charged is the total number of full calendar days on which the injured person was unable to work as a result of the injury. The total does not include the day the injury occurred or the day the injured person returned to work, but it does include all intervening calendar days (including Sundays, days off, or days when the plant is shut down.) It also includes any other full days of inability to work because of the specific injury, subsequent to the injured person's return to work.

Special cases. Days charged in special cases are listed in American National Standard Z16.1 and in NSC Industrial Data Sheet 527

or *Accident Prevention Manual.*

In the course of employment

A work injury is one that arises out of and in the course of employment; in other words, results from the work activity or environment of employment.

Employment is defined in ANSI Z16.1 as:

• All work or activity performed in carrying out an assignment or request of the employer, including incidental and related activities not specifically covered by the assignment or request

• Any voluntary work or activity undertaken while on duty with the intent of benefiting the employer, or

• Any activity undertaken while on duty with the consent or approval of the employer

Interpretation of Z16.1 rates

The injury frequency rate shows the rate of occurence of disabling injuries. Since a relationship is needed to give the rate meaning, disabling injuries are related to 1,000,000 employee-hours of work. A rate of 20.0 means that disabling injuries were incurred at the rate of 20 for each 1,000,000 employee-hours worked.

The injury severity rate shows the rate at which days are lost or charged in relation to 1,000,000 employee-hours of work. A severity rate of 500 means that 500 days were lost or charged for every 1,000,000 employee-hours worked.

Average days charged per disabling injury shows how serious the injuries were on the average and, thus, may reveal conditions not readily apparent from a review of frequency or severity rates alone. It thus makes possible a more complete evaluation of injury experience.

Off-the-Job Injuries

In recent years, off-the-job disabling injuries of employees have exceeded on-the-job disabling injuries by 30 to 40 percent. Since the unscheduled absence of employees for any reason can cause production slowdowns and delays, costly retraining and replacement, or costly overtime by remaining employees, many employers are concerned with the off-the-job injuries that occur to their employees. Moreover, activity to reduce off-the-job accidents should help to promote interest in on-the-job safety.

The ANSI Z16.3 Standard *Method of Recording and Measuring the Off-the-Job Disabling Accidental Injury Experience of Employees* provides a means for recording and measuring these off-the-job injuries—those injuries suffered by an employee which do not arise out of and in the course of employment. Definitions and rates used under ANSI Z16.3 are very similar to those used under ANSI Z16.1.

Because the data on off-the-job injuries is not as easy to obtain, however, certain simplifications are introduced in ANSI Z16.3. Exposure (for use in rates per million employee-hours) is standardized at 312 employee-hours per employee per month (equal to $4^{1}/_{3}$ man-weeks less 40 hours per week at work and 56 hours per week for sleeping).

For the calculation of severity rates, each permanent partial disability is recorded at 390 days of disability (based upon Bureau of Labor Statistics' averages for all permanent partial disabilities). Provision is also made in ANSI Z16.3 for recording home, public, and transportation injuries separately to allow for concentrated effort in problem areas.

Patron Injuries

Many employers concerned with employee injuries and workers' compensation have a similar or even greater concern about injuries to customers, patients, hotel guests, diners, tenants, or others not employed by the reporting facility. For this reason, the National Safety Council initiated and sponsored American National Standard Z108.1, *Method of Measuring and Recording Patron and Non-Employee Injury Statistics.*

A reportable patron injury is defined as one that occurs to a patron in the service environment (such as a store, parking lot, hotel, office, or restaurant), or from a service-connected accident or illness resulting from the use of a service or product, which requires professional medical or dental treatment or examination. Property damage or injuries not serious enough to warrant professional attention (or which upon examination reveal no actual injury) need not be recorded. Accidents, involving street and highway vehicles and passenger-carrying operations that are subject to federal reporting systems, would also be excluded.

Since the exposure base would vary widely between services such as theaters, with fairly consistent exposure time, and bus terminals, with inconsistent exposure, ANSI Z108.1 provides that the trade association or other national representative group determine its exposure base. In other words, the patron injury frequency rate may use as an exposure base 10,000 sales transactions, 100,000 patron-days, $1 million gross sales, or other common denominators.

The measure of severity (patron loss injury rate) is also chosen by the industry or business group. It might be the dollar value of incurred losses (claims) per $1 million sales or some other basis, generally over a twelve-month time span.

Special Records and Reports

Medical records

Medical surveillance. Under the existing OSHA surveillance programs, the employer is required to provide medical examinations for employees prior to their employment in certain processes and operations which are known to be hazardous to health, and at specified intervals during their period of employment therein, and at the termination of the employee's services.

Maintenance of records. It is imperative that the employer establish a meticulous system for maintaining all such medical records since they must be available on demand by government officials, employees, private physicians in behalf of exposed employees, and by others who may have a legal right to them. The law requires that employers of employees examined under these regulations maintain complete and accurate records of all such medical examinations, and that such records shall be retained by the employer for at least 20 years, except for vinyl chloride regulations which requires retention of records for not less than 30 years. (See Table 5-B for listing of dangerous substances that require medical surveillance records.)

Access to medical records. The contents of records of the medical examinations required by the OSHA regulations must be made available, for inspection and copying, to the Assistant Secretary of Labor for Occupational Safety and Health, the Director of the National Institute for Occupational Safety and Health (NIOSH), to authorized physicians and medical consultants of either of them and, upon request of an employee or former employee, to his physician. Any physician who conducts a medical examination required by these regulations must furnish to the employer of the examined employee all the information specifically required herein, and any other medical information related to occupational health exposures.

Employee exposure and monitoring records

Maintenance of records. Every employer shall maintain records

TABLE 5-B
HAZARDOUS MATERIALS AND EXPOSURES
SPECIAL RECORDS AND REPORTING REQUIREMENTS

		29 CFR 1910 - Subsections			
		Col. 1	*Col. 2*	*Col. 3*	*Col. 4*
Asbestos	1910.1001	(j)	—	(i)	—
4-Nitrobiphenyl	1910.1003	(g)	(f)	—	(d)
alpha-Naphthylamine	1910.1004	(g)	(f)	—	(d)
4,4'-Methylene bis(2-chloroaniline)	1910.1005	(g)	(f)	—	(d)
Methyl chloromethyl ether	1910.1006	(g)	(f)	—	(d)
3,3'-Dichlorobenzidine (and its salts)	1910.1007	(g)	(f)	—	(d)
bis-Chloromethyl ether	1910.1008	(g)	(f)	—	(d)
beta-Naphthylamine	1910.1009	(g)	(f)	—	(d)
Benzidine	1910.1010	(g)	(f)	—	(d)
4-Aminodiphenyl	1910.1011	(g)	(f)	—	(d)
Ethyleneimine	1910.1012	(g)	(f)	—	(d)
beta-Propiolactone	1910.1013	(g)	(f)	—	(d)
2-Acetylaminofluorene	1910.1014	(g)	(f)	—	(d)
4-Dimethylaminoazobenzene	1910.1015	(g)	(f)	—	(d)
N-Nitrosodimethylamine	1910.1016	(g)	(f)	—	(d)
Vinyl chloride	1910.1017*	(m)	(n)	(m)	(m)
Ionizing radiation	1910.96	—	(1)&(m)	(d)&(n)	—

Column 1. Medical surveillance and medical record requirements

Column 2. Requirements for special operational and incident reports

Column 3. Requirements for employee exposure and environmental records

Column 4. Requirements for establishing and maintaining daily reports for employees entering regulated area

1910.1017 requires retention of records for not less than 30 years; others require retention for at least 20 years.

of any personal or environmental monitoring required by the regulations covering materials listed in Table 5-B. Records shall be maintained for a period of at least 3 years, and retained for longer periods when specifically required such as vinyl chloride, where retention is for 30 years. These records must be made available upon request to the Assistant Secretary of Labor for Occupational Safety and Health, the Director of NIOSH, and to authorized representatives of either.

Employee access to records. Every employee and former employee shall have reasonable access to any record required to be maintained by the above paragraph, which indicates the employee's own exposure to hazardous substances listed in Table 5-B.

Employee notification. The importance of maintaining accurate records can not be overstressed. Since the law requires that any employee found to have been exposed at any time to concentrations of dangerous materials listed in Table 5-B in excess of the limits prescribed in the requirements shall be notified in writing of the exposure as soon as practicable but not later than 5 days of the finding. The employee shall also be timely notified of the corrective action being taken.

Incident reports

Notifications. The employer is required to notify appropriate officials of any employee over-exposure or excessive leaks of dangerous materials listed in Table 5-B, within the prescribed time limits. In the case of ionizing radiation, *immediate* notification is required in any case of radiation over-exposure and damage to property in excess of $10,000.

Reports. The regulations covering the dangerous substances listed in Table 5-B, also require the submission of special incident reports, particularly in cases of over-exposure to ionizing radiation and the carcinogens listed above.

Daily rosters and retention, access to records

Employee identification. In operations where the carcinogens listed in Table 5-B are used the employer must maintain a daily roster of employees entering regulated areas. The rosters or a summary of the rosters shall be retained for a period of 20 years.

Access to records. The rosters and/or summaries shall be provided upon request to authorized representatives of the Assistant Secretary of Labor and the Director of NIOSH. In the event that the employer ceases business without a successor, rosters shall be forwarded by registered mail to the Director.

Accident Investigation

Successful accident prevention requires a minimum of four fundamental activities:

• All working areas are studied to detect and eliminate or control physical or evironmental hazards which contribute to accidents.

• All operating methods and practices are studied.

• Education, instruction, training, and discipline are utilized to minimize human factors which contribute to accidents.

• For cause analyses, a thorough investigation of at least every accident which results in a disabling injury (under ANSI Z16.1) or OSHA lost workdays to determine contributing circumstances (see the previous chapter). Accidents that do not result in personal injury (so-called "near-accidents" or "near-misses") are warnings. They should also be investigated, but the amount of time spent should vary with the circumstances.

This fourth activity, accident investigation and analysis, is a

defense against hazards that (*a*) are overlooked in the first three activities, (*b*) are not obvious, or (*c*) are the result of combinations of circumstances that are difficult to foresee.

Why Accidents Must Be Investigated

Accident investigation and analysis is one of the means to prevent accidents. As such, the investigation or analysis must produce information that leads to countermeasures which prevent or reduce the number of accidents. The more complete the information, the easier it will be for the individual responsible for safety to design effective counter-measures. For example, knowing that 40 percent of a firm's accidents involve ladders is not as useful as knowing that 80 percent of the ladder accidents involve broken rungs.

A good recordkeeping system, as outlined in Chapter 5, is essential to accident investigation because it allows the basic facts about an accident to be recorded

Figure 6-1.—Supervisors should investigate every accident, including those not resulting in personal injury. A tape recorder speeds the work.

quickly, efficiently, and uniformly. An investigation of at least every ANSI Z16.1 disabling injury or every OSHA lost workday case should be made Accidents resulting in nondisabling injuries or no injuries (and also near-accidents) should be investigated because a "near-miss" could have been a fatality, especially if there is frequent recurrence of certain types of nondisabling injuries, or if the frequency of accidents is high in certain areas or operations.

Fact finding, not fault finding

For purposes of accident prevention, investigations must be fact-finding, not fault-finding; otherwise, they may do more harm than good. This is not to say that responsibility may not be fixed where personal failure has caused injury, or that such persons should be excused from the consequences of their actions. What this does mean is that the investigation itself should be concerned only with the facts. The investigating individual, board,

99

or committee is best kept free from involvement with the punitive aspects of their investigation.

The principal purposes of an accident investigation are:

• To learn accident causes so that similar accidents may be prevented by mechanical improvement, better supervision, employee training, or changes in procedure.

• To determine if a change in procedure produced an error that in fact resulted in an accident.

• To publicize the particular hazard among employees and their supervisors and to direct attention to accident prevention in general.

• To determine facts bearing on legal liability. An investigation undertaken only for this purpose, however, will seldom develop enough information for accident prevention purposes. On the other hand, an investigation for preventive purposes will disclose facts which are important in determining liability.

Types of Investigation Techniques

There is a variety of accident investigation and analysis techniques available. Some of these techniques are more complicated than others. The choice of a particular method will depend upon the purpose and orientation of the investigation.

The accident investigation and analysis procedures, explored here is based upon the American National Z16.2 Standard, which focuses primarily on unsafe acts and unsafe conditions and is the most often used technique. Other similar techniques involve investigation within the framework of defects in man, machine, media, and management, or in education, enforcement, and engineering.

For analysis, these techniques involve classifying the data about a group of accidents into various categories. This is known as the statistical method of analysis. Countermeasures are designed on the basis of most frequent patterns of occurrence.

Other techniques come under the systems approach to safety. "Systems safety" takes into account the interrelationships between the various events that can lead to an accident. As accidents will rarely have one cause, the systems approach to safety can point to more than one place in a system where effective countermeasures can be introduced.

Who Should Investigate

Depending on the nature of the accident and other conditions, the investigation may be made by the foreman, the person responsible for safety, the workers' safety committee, the general safety committee, or an engineer from the insurance company. If the accident involves special features, consultation with an engineer, or industrial hygienist, or other specialist might be warranted.

The foreman or supervisor

The foreman should make an immediate report of every accident; see Figure 5-4 in Chapter 5. Since he is right there, he probably knows more about the accident than anyone else, and it is up to him, in most cases, to put into effect whatever measures may be adopted to prevent similar accidents. (See Figure 6-1.)

The safety professional

If there is a safety professional on the satff, he should make an investigation of every important accident for his own information, and, in most cases, he should draft a written report to the proper management official or to the general safety committee.

Nowhere are the safety professional's value and ability better shown than in the investigation of an accident. His specialized training and analytical experience enable him to get all the facts, apparent and hidden, and to submit an unbiased report. He has no interest in the investigation other than to obtain information which can prevent a similar accident.

The workers' safety committee

In some companies, one function of a workers' safety committee is to investigate and report on accidents. This function is particularly important when a contributing factor is determined to be an unsafe act on the part of the worker. If this committee finds and reports that the injured person did engage in an unsafe act that contributed to the accident and if this report is then publicized, it will create a much deeper impression among the workers than a similar report made by the foreman or safety professional.

A general safety committee

In many companies, especially those of small or moderate size, all safety activities are headed by a general safety committee, one of whose activities is accident investigation. Ordinarily, such investigation would be handled in a routine manner, but in important cases the chairman might call an extra meeting of the committee to conduct a special investigation.

Investigation procedures

The following procedures have been found particularly effective when investigating accidents:

a. Go to the scene of the accident promptly.

b. Talk with the injured person, if possible. Talk with witnesses. Stress getting the facts, not placing the responsibility or blame.

c. Listen for clues in conversations. Unsolicited comments often have merit.

d. Encourage people to give their ideas for preventing the accident.

e. Study possible causes—both unsafe conditions and unsafe practices.

f. Confer with interested persons about possible solutions. The problem may have been solved by someone else.

g. Write up a report, using a printed form which allows a narrative description. See Figure 5-4 in Chapter 5, pp. 85-86.

h. Follow up to make sure conditions are corrected. If they cannot be corrected immediately, report this to all concerned.

i. Publicize any corrective action taken so that all may benefit from the experience.

101

When following this procedure, keep these two basic points in mind:

• That most accidents involve both unsafe conditions and unsafe acts, and

• That the purpose of accident investigation is to prevent future accidents and not to fix blame.

What Should Be Investigated

An accident that causes death or serious injury obviously should be investigated. The near-accident that *might* have caused death or serious injury is equally important from the safety standpoint and should be investigated, such near misses as the breaking of a crane hook or a scaffold rope, or the explosion of a tank.*

Each investigation should be made as soon after the accident as possible. A delay of only a few hours may permit important evidence to be destroyed or removed, intentionally or unintentionally. Also, the results of the inquiry should be made known quickly, inasmuch as their publicity value in the safety education of employees and supervisors is greatly increased by promptness.

Any epidemic of minor injuries demands study. A particle of emery in the eye or a scratch from handling sheet metal may be a very simple case—the immediate cause is obvious and the loss of time may not exceed a few minutes. However, if cases of this or any other type occur frequently, an investigation should be made to determine the underlying causes. The energetic manager is con-stantly alert to the advantage of this kind of accident investigation which may prove more valuable, though less spectacular, than the "inquest" following a fatal injury.

Fairness is absolutely essential. The value of the investigation is largely destroyed if there is any suspicion that its purpose is to place the blame or pass the buck. No one should be assigned to investigative work unless he has earned a reputation for fairness and is experienced in gathering evidence. It should be made quite clear that accident investigations are conducted entirely for the purpose of obtaining information which will help to prevent recurrence of accidents.

Identifying the Key Facts

As explained in ANSI Z16.2, the purpose of the investigation to identify certain key facts about each injury and the accident that produced it, and to record those facts in a form that will permit summarization to show general patterns of injury and accident occurrence, in as great an analytical detail as possible. These patterns are intended to serve as guides to the areas, conditions, and circumstances to which accident prevention efforts most profitably may be directed.

---

*Reports of incidents involving (a) release of carcinogenic materials required by OSHA regulations 29 CFR 1910.1003 through .1017, and (b) overexposure and/or excessive levels of ionizing radiation, required by 29 CFR 1910.96(l) and (m) may require in depth investigations. See Table 5-B, page 96.

For a complete recording of an injury case, one item for each key fact should be selected from the injury report. (Whether or not all the key facts are present in a case will be determined by the circumstances of that case.) The items should be selected according to the following definitions and must be developed in any accident investigation:

Nature of injury—the type of physical injury incurred.

Part of body—the part of the injured person's body directly affected by the injury.

Source of injury—the object, substance, exposure, or bodily motion which directly produced or inflicted the injury.

Accident type—the event which directly resulted in the injury.

Hazardous condition—the physical condition or circumstance which permitted or occasioned the occurrence of the accident type.

Agency of accident—the object, substance, or part of the premises in which the hazardous condition existed.

Agency of accident part—the specific part of the agency of accident that was hazardous.

Unsafe act—the violation of a commonly accepted safe procedure which directly permitted or occasioned the occurence of the accident event.

Often supplementary items of information closely related to the key facts, such as sex, occupation, and type of work being performed at the time the injury was incurred, are developed during an investigation and must be recorded and included in an analysis so that all the facts will be available for taking the proper preventive steps.

The principal source of information for an analysis is the supervisor's accident report (Figure 5-4, Chapter 5). Complete data of all key facts should be fully and accurately recorded on this form at the time of the accident investigation.

Injury and accident reports, however, commonly consist of a few specific statements relating to the injury, plus a narrative account of how and why the accident occurred. The reports vary widely in the amount of detail given and in the clarity and coherence with which the facts are presented. Therefore, an attempt should be made to educate those investigating accidents so that they will be sure to identify the key facts involved during the investigation. Table 6-A provides a checklist for identifying key facts during an investigation.

Examples of identifying key facts. The first step in analyzing an injury or accident report for statistical purposes consists of selecting from the report the answers to the following questions. (See Table 6-A.)

1. *Nature of injury.* What was the injury?

2. *Part of body.* What part of the body was affected by the injury named in (1)?

3. *Source of injury.* What object, substance, exposure, or bodily motion inflicted the injury?

TABLE 6-A
CHECKLIST FOR IDENTIFYING KEY FACTS

1. NATURE OF INJURY

Foreign body	Strain and sprain	Amputation	Dermatitis
Cut	Fracture	Puncture wound	Ganglion
Bruises and contusions	Burns	Hernia	Abrasions
			Others......................

2. PART OF BODY

Head and Neck	Upper Extremities	Body	Lower Extremities
Scalp	Shoulder	Back	Hips
Eyes	Arms (Upper)	Chest	Thigh
Ears	Elbow	Abdomen	Legs
Mouth, teeth	Forearm	Groin	Knee
Neck	Wrist	Others......................	Ankle
Face	Hand		Feet
Skull	Fingers and thumb		Toes
Others......................	Others......................		Others......................

4. ACCIDENT TYPE

Struck against (rough or sharp objects, surfaces etc. exclusive of falls)	Struck by sliding, falling or other moving objects	Overexertion (resulting in strain, hernia, etc.)	Inhalation, absorption, ingestion, poisoning, etc.
Struck by flying objects	Caught in (on or between)	Slip (not a fall)	Contact with electric current
	Fall on same level	Contact with temperature extremes, burns	Others......................
	Fall to different level		

5. HAZARDOUS CONDITION

Improperly or inadequately guarded	Defective tools, equipment, substances	Hazardous arrangement	Poor housekeeping
Unguarded	Unsafe design or construction	Improper illumination	Congested area
		Improper ventilation	Others......................
		Improper dress	No unsafe condition

6. AGENCY OF ACCIDENT

Machine	Can and end conveyors (belt, cable, can dividers, chain, twisters, drops, can elevators, etc.)	Hoists and Cranes	Chemicals
Vehicles		Elevators (passenger and freight)	Ladders or scaffolds
Hand tools		Building (door, pillar, wall, window, etc.)	Electrical apparatus
Tin and black plate (sheet, stock, or scrap)	Conveyors (chutes, belt, gravity)	Floors or level surfaces	Boilers, pressure vessels
Material work handled (other than tin and black plate)		Stairs, steps, or platforms	Others......................

8. UNSAFE ACT

Operating without authority	Using equipment, tools, materials or vehicles unsafely	Unsafe loading, placing and mixing	Adjusting, clearing jams, cleaning machinery in motion
Failure to warn or secure		Unsafe lifting and carrying (including insecure grip)	Distracting, teasing
Operating at unsafe speed	Failure to use personal protective equipment		Poor housekeeping
Making safety devices inoperative	Failure to use equipment provided, (except personal protective equipment)	Taking an unsafe position	Others......................
Using defective equipment, materials, tools or vehicles			No unsafe act

CONTRIBUTING FACTORS

Disregard of instructions	Lack of knowledge or skill	Failure to report to medical department	Others......................
Bodily defects	Act of other than injured		No contributing factor

4. *Accident type.* How did the injured person come in contact with the object, substance, or exposure named in (3), or during what personal movement did the bodily motion named in (3) occur?

5. *Hazardous condition.* What hazardous physical or environmental condition or circumstance caused or permitted the occurrence of the event named in (4)?

6. *Agency of accident.* In what object, substance, or part of the premises did the hazardous physical or environmental condition named in (5) exist?

7. *Agency of accident part.* In what specific part of the agency of accident named in (6) did the hazardous condition named in (5) exist?

8. *Unsafe act.* What unsafe act of a person caused or permitted the occurrence of the event named in (4)? Contributing factors should be noted when they can be determined.

Table 6-A gives typical checklists that can be used when making the report. For easier statistical handling of the data, the factors can be individually numbered, and the numbers inserted on the form. (See Classifying the Key Facts, discussed on page 106.)

The following examples show how to identify the key facts in an injury case for purposes of classification.

ACCIDENT 1. The operator of a circular saw reached over the running saw to pick up a piece of scrap. His hand touched the blade, which was not covered, and his thumb was severely lacerated.

1. Nature of injury—laceration
2. Part of body—thumb
3. Source of injury—circular saw
4. Accident type—struck against
5. Hazardous condition—unguarded
6. Agency of accident—circular saw
7. Agency of accident part—saw teeth
8. Unsafe act—cleaning moving machine

ACCIDENT 2. A lift truck went out of control when one wheel hit a piece of stock lumber which projected into the aisle. The truck ran out of the aisle and struck a machine operator, breaking his left leg between the ankle and knee.

1. Nature of injury—fracture
2. Part of body—lower leg
3. Source of injury—lift truck
4. Accident type—struck by
5. Hazardous condition—improperly placed lumber
6. Agency of accident—lumber
7. Agency of accident part—none
8. Unsafe act—unsafe placement of material

ACCIDENT 3. A warehouse employee jumped from the loading platform to the ground instead of using the steps. As he landed, he sprained his right ankle.

1. Nature of injury—sprain
2. Part of body—ankle
3. Source of injury—ground
4. Accident type—fall from elevation

105

5. Hazardous condition—none indicated
6. Agency of accident—none
7. Agency of accident part—none indicated
8. Unsafe act—jumping down.

ACCIDENT 4. A laborer working in a trench was suffocated under a mass of earth when the unshored wall of the trench caved in.
1. Nature of injury—asphyxia
2. Part of body—respiratory system
3. Source of injury—earth
4. Accident type—caught under
5. Hazardous condition—lack of shoring
6. Agency of accident—trench
7. Agency of accident part—none
8. Unsafe act—none indicated

ACCIDENT 5. A salesman drove his car through an intersection with the green traffic light in his favor. He was struck by another car coming from his right that ran the red light. The whiplash effect fractured a vertebra in the salesman's neck.
1. Nature of injury—fracture
2. Part of body—neck
3. Source of injury—auto
4. Accident type—collision
5. Hazardous condition—traffic violation
6. Agency of accident—vehicle
7. Agency of accident part—none
8. Unsafe act—none (by the salesman)

Classifying the Key Facts.

Even in large company operations in which hundreds of acci-

dents may occur annually, only rarely do two accidents occur in exactly the same way. Accidents do follow general patterns, however, and grouping them according to pattern is necessary for analysis.

Setting up classifications

Before the actual analysis is begun, classifications must be devised for grouping the various data. For each key fact, general classifications should be established in which similar data may be grouped. Then, more specific classifications should be set up within each general classification to preserve as many of the details as possible.

For example, in American National Standard Z16.2, among the general classifications recommended for the key fact HAZARDOUS CONDITION are the following:

a. Defects of agencies
b. Dress or apparel hazards
c. Environmental hazards
d. Placement hazards
e. Inadequate guarding
f. Public hazards

Within each of one of these general classifications, more specific classifications are set up. Under DEFECTS OF AGENCIES, for example, are listed:

a. Composed of unsuitable materials
b. Dull
c. Improperly constructed, assembled, etc.
d. Improperly designed
e. Rough
f. Sharp

g. Slippery

h. Worn, cracked, broken, etc.

i. Other

It is not always possible to set up classifications before the analysis is begun. In this case, classifications can be developed as reports are reviewed and situations are revealed.

For example, if an analysis is being made of ladder accidents and it is found that in a number of cases broken rungs caused the accidents, a specific group, *Broken Rungs,* should be set up under the classification DEFECTS OF AGENCIES.

Standard Z16.2 recommends general and specific classifications for all the key facts. These classifications are presented principally as suggestions to guide the individual in devising classifications to fit his own problems. The point must be emphasized that for an analysis to be of maximum usefullness, classifications must encompass the situations which are pertinent to the particular company.

Use of a numerical code

Regardless of the method which eventually will be used to sort and tabulate the various key facts, the work will be facilitated if code numbers are assigned to the different classifications. With this method, each case need be read only once, at which time code numbers are assigned to the different facts, and subsequent sorting of the various facts can be quickly completed merely by reference to the code numbers.

A numerical code is simply the assigning of numbers in sequence to a list of similar facts. For each key fact (Agency of accident, Accident type, Hazardous condition, etc.), there should be no duplication of numbers, but for the different facts, a subset of numbers should be used. Figure 6-2 shows how code numbers were assigned for two key facts in an analysis of ladder accidents.

After the cases have been reviewed and code numbers have been assigned to the different key facts, the reports can be easily and quickly sorted or arranged by any of the facts to reveal the principal data concerning the accidents.

Numerical codes are already assigned to all the classifications which are included in Standard Z16.2, and if the individual uses this standard, he can use these code numbers, too. If he uses the standard as a starting point and adds other classifications to cover the specific accident experience of his own company, he can code the additional items to fit into the code of the standard.

Accident Analysis

Accident prevention has progressed from a hit-or-miss proposition to a technique closely approaching the scientific method. Whereas, in years past, a reduction in accident rates was attempted primarily by humanitarian appeal to management and workers, today methods are employed that isolate and identify the causes of accidents and that permit direct and positive action to be taken to prevent their recurrence.

Like other phases of modern business management, accident

Analysis of Ladder Accidents

Code No.	Unsafe Conditions		Total no. of cases
1	Slippery rungs	THL I	6
2	Broken rungs	III	3
3	Lack of hooks for fastening ladder at top	I	1
4	Worn ladder shoes	I	1
5	Weak, worn, cracked rails	II	2

Unsafe Acts

10	Failure to secure ladder at top	THL THL THL THL THL IIII	29
20	Failure to secure ladder at bottom	THL THL II	12
30	Placing ladder unsafely		22
31	On boxes or other equipment	THL THL IIII	14
32	Near moving equipment	II	2
33	On planking over opening	I	1
34	On inclined or irregular surface	II	2
35	At too great an angle	III	3
40	Working in unsafe position		10
41	Overreaching	THL I	6
42	Improper stance while using leverage tools	II	2
43	Stradling space between ladder and nearby object	II	2
50	Ascending or descending improperly		35
51	Improper or insecure grip	THL IIII	9
52	With back to ladder	I	1
53	Running up or down ladders	THL I	6
54	Jumping from lower steps to ground	THL	5
55	Oily or wet shoes	III	3
56	Carrying too heavy loads	THL THL I	11
60	Using ladders known to be defective	THL III	8

Figure 6-2.—Tabulations of unsafe conditions and unsafe acts which contributed to ladder accidents. The information was taken from the supervisor's accident reports and based the classifications upon the actual experiences as reported.

prevention must be based on facts that clearly identify the problem. An approach to the accident prevention problem on this basis will not only result in more effective control over accidents, but also will permit this objective to be accomplished with savings in time, effort, and money.

Experience has proved that the most effective way to reduce accidents is to concentrate on one phase of the accident problem at a time rather than try to stop all accidents at once. There are different ways in which the problem can be approached on this basis, any one of which should prove effective.

The reports may be grouped by occupation of the injured person. Each group of reports, then, may be reviewed in order to determine what accident types, sources of injury, and agencies of accident are most prevalent among different occupations. Such information is particularly helpful in planning employee training and in developing educational materials and programs.

Injury frequency rates computed by department or operation may reveal that injuries occur at sharply higher rates in some departments or operations than in others. If such is the case, an analysis should be made of the accidents in the high-rate departments to find the sources of the accidents and the hazardous conditions and agencies that caused them. This method will permit concentration of effort in the locations in which accidents occur most frequently.

If injury frequency rates reveal that a high rate of occurrence is general throughout the operation, the accident reports might be grouped by agency of accident, source of injury, accident type, unsafe act, or hazardous condition. Any one of the key facts may be a starting point for analysis.

After injury rates have been used to identify the operations or departments in which injuries occur most frequently, analysis of individual cases will provide the information necessary to reduce accidents in these locations. Sometimes a high company rate is not identified with one or a few departments, but instead represents a high occurrence of accidents throughout the company. Under such circumstances, it is even more important that an analysis of the accidents be made.

Similar accidents may occur frequently but at widely separated locations, so that their high incidence is not always apparent. Accidents may be more numerous in the performance of certain repetitive tasks. Some unsafe practices which cause accidents may be committed repeatedly but at different times and in different places, so that their importance as accident causes is not immediately recognized.

Results From Good Accident Investigation and Analysis

Analysis of the circumstances of accidents, using ANSI Z16.1 and Z16.2, can produce these results:

• Identity and location of the principal sources of accidents by determining, from actual experiences, the materials, machines, and tools most frequently involved in accidents and the jobs most likely to produce injuries

• Disclosure of the nature and size of the accident problem in departments and among occupations

109

- Indication of the need for engineering revision by identifying the principal unsafe conditions of various types of equipment and materials

- Disclosure of inefficiencies in operating processes and procedures where poor layout, for example, contributes to accidents, or where outdated methods or procedures which overtax the physical capacities of the workers can be avoided, as by using mechanical handling methods

- Disclosure of the unsafe practices which need special attention in the training of employees

- Disclosure of improper placement of personnel in instances in which inabilities or physical handicaps contribute to accidents

- Use by supervisors of the time available for safety work to the greatest advantage by providing them with information about the principal hazards and unsafe practices in their departments

- The objective evaluation of the progress of a safety program by noting in continuing analyses the effect of different safety measures, educational techniques, and other methods adopted to prevent injuries

- Compliance with federal and state requirements

Measuring Safety Performance By Accident and Cost Analysis

Of course, merely obtaining the information will not prevent recurrence of accidents. The conditions which contributed to the accidents must be corrected.

A sample accident analysis

An analysis "by agency of accident" will always reveal information which can be used effectively in reducing accidents. In one company, a grouping of accidents in this manner revealed a large number involving ladders and, because accidents of this type often resulted in serious injuries, these cases were analyzed in order to obtain information that could be used to decrease their occurrence.

Separate sheets were used to list the classifications which were developed for each fact. Also shown were the code numbers assigned these classifications and a tally of occurrences.

Figure 6-2 shows the sheet which was developed for hazardous conditions and unsafe acts. Each classification is specific, relating directly the ladder accidents. Similar sheets were developed for accident type and nature of injury.

After each accident report was reviewed, the proper code number was placed opposite each fact on the report.

Because this analysis of ladder accidents was relatively simple, single classifications of each fact revealed practically all the information necessary to correct the situations which contributed to the accidents.

The causes of the ladder accidents were ascertained in terms so definite that guesswork could be completely replaced by direct, positive action, which could have no other result than a marked decrease in the frequency of this kind of accident.

A sample accident cost analysis

Because they recognize accident prevention as a significant part of the supervisor's job, some companies set up their accounting procedures to record medical and compensation expenses by department. Some even list supervisor's names, with accident costs incurred by their respective departments, in monthly or annual reports to top management.

So that valid comparisons of accident cost can be made between departments (and companies and industries), the exposures must be similar. Standard cost factors are usually based on cost per 1000 man-hours worked, or premium per $100 of payroll under workers' compensation.

Departmental accident cost may be classified by source of injury, such as portable electric tools or handling materials. This type of classification system permits supervisors to find their most expensive trouble spots.

Accident cost may be classified another way—by listing the cost as either insured or uninsured.

Insured costs include compensation insurance premiums and, in some cases, medical expenses. These costs are easy to obtain.

Uninsured costs are more difficult to determine, but are estimated by the National Safety Council to be about equal to (and sometimes higher than) the insured or visible costs. Those costs that can be determined should be placed on the Supervisor's Accident Cost Report (see Figure 6-3).

A cost evaluation can usually be made on the following items: production loss, time lost by supervisors, time lost by co-workers of injured employee, and losses from damaged materials or machines. Each is discussed in turn.

● Production losses come from partial or complete shutdown due to:

1. Damage of material, machines, or production area.
2. Emotional upset and lowered employee morale which tends to lower production.
3. Increased tension resulting in an increase of materials spoiled and pieces rejected.
4. Replacement of employee(s) who produce less while being trained to fill injured worker's job.

● Time lost by supervisors can be due to:

1. Assisting injured employee.
2. Investigating the accident.
3. Preparing accident reports.
4. Hiring and training new employee(s).
5. Attending hearings conducted by the state.

● Time lost by co-workers of injured employee(s):

1. In aiding injured person(s).
2. Because of curiosity.
3. Because of sympathy.

● Losses from damaged material or machines due to:

1. Cost of repair of building, machines, or tools.
2. Cost of damaged or spoiled pieces.
3. Cost (in wages paid to injured employee) because he is not fully productive.

DEPARTMENT SUPERVISOR'S ACCIDENT COST REPORT

Injury Accident **Minor injury**

No-Injury Accident_____

Date **Nov. 30, 19--** Name of injured worker **John Smith**

1. How many other workers (not injured) lost time because they were talking, watching, helping at accident? __**3**__

   About how much time did most of them lose? _____ hours __**15**__ minutes **each**

2. How many other workers (not injured) lost time because they lacked equipment damaged in the accident or because they needed the output or aid of the injured worker? __**2**__

   About how much time did most of them lose? _____ hours __**40**__ minutes **each**

3. Describe the damage to material or equipment **Unsafe sling allowed crate to fall while being hoisted. Crate and separators broke.**

   Estimate the cost of repair or replacement of above material or equipment $__**175.00**__

4. How much time did injured worker lose on day of injury for which he was paid? __**6 hours**__ hours _____ minutes

5. If operations or machines were made idle: Will overtime work probably be necessary to make up lost production? Yes ☒, No ☐. Will it be impossible to make up loss of use of machines or equipment? Yes ☐, No ☒.

   Demurrage or other special non-wage costs due to stopping an operation $__**75.00**__

6. How much of supervisor's time was used assisting, investigating, reporting, assigning work, training or instructing a substitute, or making other adjustments __**4½**__ hours _____ minutes.

   Name of supervisor ___*[signature]*___

Fill in and send to the safety department not later than day after accident.

Published by National Safety Council
425 North Michigan Avenue
Chicago, Illinois 60611

Form IS7-Rep. 2M67299   Printed in U.S.A.   Stock No. 129.27

Figure 6-3.—Accident cost report form (8½ × 11 in.) should be prepared by the department supervisor as soon after the accident as information becomes available on the amount of time lost by all persons and the extent of damage to product and equipment.

- Other losses:
1. Estimate of lost business because of late deliveries.
2. Intangible loss of good will and prestige.
3. Grievances, cost of impairment of employer-employee relationship.

From a review of the Department Supervisor's Accident Cost Report, it is a relatively easy matter to compare the direct or insured cost and the indirect or uninsured cost of an accident. If, for example, the injured worker received a minor injury, which required treatment by a doctor at a cost of $15, the direct or insured cost is $15. Assuming that the employees are paid $4.00 an hour and supervisors $6.00 an hour, the costs can be computed as follows:

Three workers lost time because they were either talking, watching, or helping at the scene of the accident. Each of these workers lost 15 minutes. Two other workers lost time because they lacked equipment damaged in the accident or because they needed the outpout or help of the injured worker. Both of these workers lost approximately 40 minutes each. The injured worker lost a total of six hours; he was paid his full wages during this period. The supervisor's time used in assisting, investigating, reporting, assigning work, training or instructing a substitute or making adjustments, amounted to $4\frac{1}{2}$ hours. Therefore, a labor cost of $59.33 is developed. The estimated cost of repair or replacement of the damaged equipment is approximately $175.00. Overtime work is necessary to make up for the lost production and will cost $75.00. Therefore, the direct and indirect cost of this minor accident is:

DIRECT COST

| Payment to the doctor | $ 15.00 |

INDIRECT COST

Labor cost	$ 59.33
Repair or replacement	175.00
Overtime costs	75.00
Total indirect (or hidden) cost	$309.33
Direct cost of the above	$ 15.00
Hidden cost of the above	309.33
TOTAL COST	$324.33

Training

People do unsafe things because:

- They are unaware that what they are doing is wrong.

- They misunderstand instructions that are given.

- They do not consider the instructions to be important.

- They are not given specific instructions.

- They find it awkward to follow the instructions.

- They deliberately disregard instructions.

Often they are unaware that what they are doing is wrong because they lack the knowledge or skill to do their job properly and safely. This is one of the most frequent personal causes of accidents, and may be due to one of these reasons:

- The person may never have learned to do his job the right way.

- He may have learned it once, but not well enough for correct, safe, work habits to stick with him.

- He may have learned how to do the job safely under normal conditions, but has never fully realized the danger of certain unsafe acts, possibly involving some unusual job conditions.

Good instruction is important to safety. This section discusses: how to give instruction, the need and importance of job safety analysis (JSA), methods of making a job safety analysis, and job instruction training.

Who Instructs?

Sometimes a supervisor gives job instruction, but often it is someone else who does the teaching. The supervisor is frequently too busy with other work to be able to instruct every trainee. In fact somebody else—the boss, an assistant foreman or an experienced worker for example—may have more firsthand knowledge of the job than does the supervisor.

Regardless of who does the breaking-in job, it is still the responsibility of the supervisor (or person in charge) to know what is being taught and whether or not the trainee is really mastering it. The supervisor should occasionally observe the instructor at work, and check each trainee's progress. This keeps him informed, and shows his interest as well. The student must feel that whoever gives the instruction, the person in charge is responsible for whatever is taught.

Training the New Employee

At the time he is employed, even before he starts his job, the new employee should be told something about company and department safety policies. This can be brief, but must be the good start that will make later instruction and supervision easier.

A new employee wants to get started on his job. He can be worn down by too much talk, much of which he cannot understand until he has seen what the job is like. No matter how brief this initial talk, some things must be said. The person giving the talk should think about them carefully.

• Management and employees are determined to prevent accidents, because accidents are bad for all concerned.

• Most employees here have never suffered a disabling injury at work, except in rare instances.

• In spite of everything the company has done, people sometimes get hurt. To avoid getting hurt, each person must look out for his own safety.

• Job instruction will include safety instruction. (No one is expected to do a job he does not understand. People are urged to ask questions about any part of their work they do not understand, and supervisors will answer questions willingly.)

• Instruction is not fault finding; correction is not reprimanding.

• It is positively forbidden for any person to try to operate any mechanical equipment without instruction and specific authority from his supervisor.

• Persons are urged to report anything about the work that seems unsafe.

• All injuries must be reported to the supervisor or person in charge.

• Methods of procuring protective equipment, and rules and practices governing its use, will be explained.

Often the employer may talk to new employees about safety before they are directed to their jobs. This is good practice. It shows management interest in safety and strongly supports the safety instruction given later.

On-the-Job Training

On-the-job training (OJT) is widely used because the trainee can be producing while he is being trained. Regardless of who does the teaching, the training should be carefully planned and organized.

In too many cases, on-the-job training is a "hit or miss" procedure where the trainee is told to follow another worker around and learn his job. In situations of this type, the worker may be too busy to do any training, or may

115

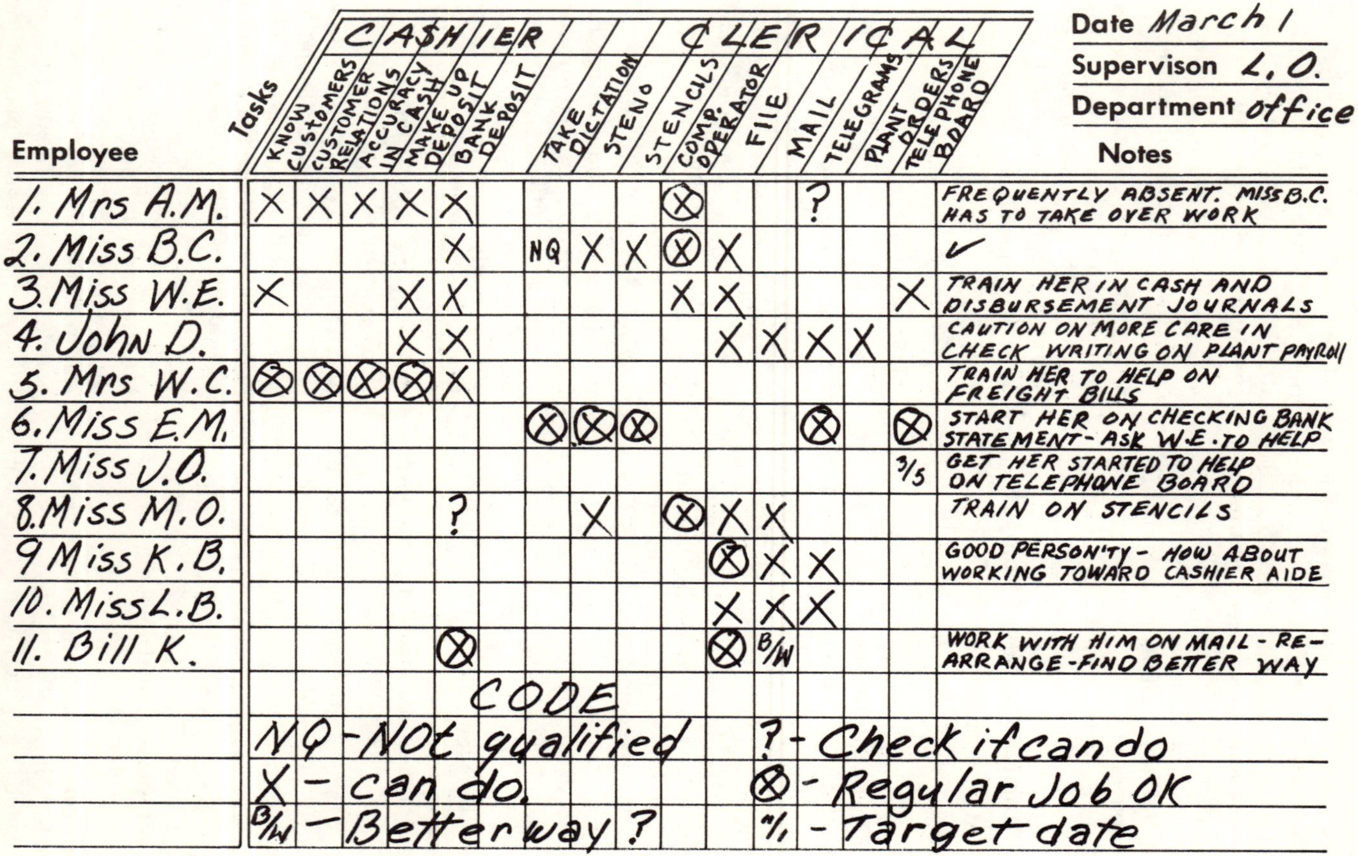

Training chart — CASHIER / STENO / CLERICAL

Employee	Know Customers	Customer Relations	Accuracy in Cash	Make up Cash	Bank Deposit	Take Dictation	Steno	Stencils	Comp. Operator	File	Mail	Telegrams	Plant Orders	Tele Phone Board	Notes
1. Mrs A.M.	X	X	X	X	X						?				Frequently absent: Miss B.C. has to take over work
2. Miss B.C.	X	X	X	X	X	NQ		⊗						✓	
3. Miss W.E.	X	X	X	X	X	X	X	X	X			X	X		Train her in cash and disbursement journals
4. John D.		⊗	⊗	X	X			X	X	X	X	X			Caution on more care in check writing on plant payroll
5. Mrs W.C.	⊗	⊗	⊗	⊗	⊗										Train her to help on freight bills
6. Miss E.M.						⊗	⊗	⊗	⊗		⊗		⊗		Start her on checking bank statement — ask W.E. to help
7. Miss J.O.					?	X		⊗					3/5		Get her started to help on telephone board
8. Miss M.O.									⊗	X					Train on stencils
9. Miss K.B.					⊗					X	X				Good person — how about working toward cashier aide
10. Miss L.B.									X	X					
11. Bill K.	B/W								⊗	B/W	B/W				Work with him on mail — re-arrange — find better way

CODE
NQ — Not qualified
X — can do
B/W — Better way?
? — Check it/can do
⊗ — Regular Job OK
3/5 — Target date

Figure 7-1.—Training chart keeps track of who needs (and who has) what training.

not know good training procedures. He may even be reluctant to train another to do his work.

Training on the job includes many techniques and approaches and there is no one method that will fit all situations. Job safety analysis (JSA) is one method, job instruction training (JIT) is another, and over-the-shoulder coaching is still another widely used method. These methods may be used separately or in combination depending upon the complexity of the job and the time element.

Over-the-shoulder coaching

Over-the-shoulder coaching is perhaps the most flexible and direct of the three training methods. In coaching, the trainee is expected to develop and apply his skills in typical work situations under the guidance of a qualified person. The person to whom the trainee has been assigned should be one who knows the job thoroughly, is a safe worker, and has the patience, time, and desire to help others.

The advantages of training of this type are:

1. The worker is more likely to be highly motivated because the guidance is personal.

2. The instructor can identify specific performance deficiencies and take immediate and proper corrective action.

3. Results of the training are readily apparent since real equipment is being used and finished work can be judged by existing standards.

4. The training is practical and realistic and can be applied at the proper time.

Timing is important; not only do the trainees like to get help when needed, but also the instructor can judge the trainee's progress continually so he can present the next unit or phase of instruction when the trainee is ready.

To help keep track of the progress of each individual, a training chart is valuable (Figure 7-1). On the top of the chart are listed the various tasks required of a person in a particular job classification. By observation and direct discussions, the instructor can determine whether the employee is qualified to perform the tasks necessary to fill the job. The instructor determines the degree of skill and knowledge of the trainee and estimates his training needs. He makes notes and comments on the chart. Such a chart prevents neglecting training that is important; it also prevents unnecessary training.

Job safety analysis

A job safety analysis should be prepared for every job where workers need to be trained in skilled operations. When there are hazards inherent in the job due to the machines or equipment used, the material worked, or the method or process used, these hazards should be indicated in the analysis and the job analysis becomes a job safety analysis, see Figure 7-2.

An experienced person should prepare the job safety analysis.

A job is a sequence of separate steps or activities that when put together accomplish a work goal.

117

JOB SAFETY ANALYSIS TRAINING GUIDE	JOB:			DATE:
	TITLE OF MAN WHO DOES JOB:		FOREMAN/SUPR:	ANALYSIS BY:
DEPARTMENT:		SECTION:		REVIEWED BY:
REQUIRED AND/OR RECOMMENDED PERSONAL PROTECTIVE EQUIPMENT:				APPROVED BY:

SEQUENCE OF BASIC JOB STEPS	POTENTIAL ACCIDENTS OR HAZARDS	RECOMMENDED SAFE JOB PROCEDURE
Break the job down into its basic steps, e.g., what is done first, what is done next, and so on. You can do this by 1) observing the job, 2) discussing it with the operator, 3) drawing on your knowledge of the job, or 4) a combination of the three. Record the job steps in their normal order of occurrence. Describe what is done, not the details of how it is done. Usually three or four words are sufficient to describe each basic job step.

For example, the first basic job step in using a pressurized water fire extinguisher would be:

1) Remove the extinguisher from the wall bracket. | For each job step, ask yourself what accidents could happen to the man doing the job step. You can get the answers by 1) observing the job, 2) discussing it with the operator, 3) recalling past accidents, or 4) a combination of the three. Ask yourself: can he be struck by or contacted by anything; can he strike against or come in contact with anything; can he be caught in, on, or between anything; can he fall; can he overexert; is he exposed to anything injurious such as gas, radiation, welding rays, etc.? for example, acid burns, fumes. | For each potential accident or hazard, ask yourself how should the man do the job step to avoid the potential accident, or what should he do or not do to avoid the accident. You can get your answers by 1) observing the job for leads, 2) discussing precautions with experienced job operators, 3) drawing on your experience, or 4) a combination of the three. Be sure to describe specifically the precautions a man must take. Don't leave out important details. Number each separate recommended precaution with the same number you gave the potential accident (see center column) that the precaution seeks to avoid. Use simple do or don't statements to explain recommended precautions as if you were talking to the man.

For example: "Lift with your legs, not your back." Avoid such generalities as "Be careful," "Be alert," "Take caution," etc. |

Figure 7-2.—Job safety analysis training guide.

Identifying hazards and stressing safe procedures are part of job safety analysis and should be one of the first steps taken where there is a possibility of injury to the worker.

Here are the steps to follow when analyzing a job:

Selecting the job. Jobs suitable for a job safety analysis (JSA) are those assignments that are given to a worker. Operating a machine, tapping a furnace, piling warehouse goods are all subjects for (JSA). They are neither too broad nor too narrow.

Jobs should not be selected at random—those with the worst accident experience should be analyzed first if JSA is to yield the quickest possible return. In fact, small companies may make this the focal point of their accident prevention program.

Break the job down. Before the search for hazards begins, a job should be broken down into a sequence of steps, each describing what is being done. Avoid the two common errors: *(a)* making the breakdown so detailed that an unnecessarily large number of steps results, or *(b)* making the job breakdown so general, that basic steps are not recorded.

The technique of making a job safety analysis involves these steps:

1. Selecting the right man to observe
2. Briefing him on the purpose
3. Observing him perform the job and trying to break it into basic steps
4. Recording each step in the breakdown
5. Checking the breakdown with the man who was observed

Select the person who is experienced, capable, cooperative, and willing to share ideas. This person will be easy to work with.

If the person has never helped on a job safety analysis, explain the purpose—to make a job safe by identifying hazards and eliminating or controlling them—and show him a completed JSA. Reassure him that he was selected because of his experience and capability.

To determine the basic job steps, ask "What step starts the job?" Then, "What is the next basic step?" And so on.

To record the breakdown, number the job steps consecutively as illustrated in the first column of the JSA training guide. Each step tells what is done, not how.

The wording for each step should begin with an action word, like "remove," "open," or "weld." The action is completed by naming the item to which the action (expressed by the verb) applied, for example, "remove extinguisher," "carry to fire."

In checking the breakdown with the person observed, an agreement of what is done and the order of steps should be obtained. The person should be thanked for his cooperation.

Identify hazards and potential accidents. Before filling in the next two columns of the JSA—Potential Accidents or Hazards and Recommended Safe Job Procedure—begin the search for hazards. The purpose is to identify

JOB SAFETY ANALYSIS TRAINING GUIDE	JOB: VERTICAL LATHE OPERATION		DATE: AUGUST 29, 19 --
	TITLE OF MAN WHO DOES JOB: LATHE MACHINIST	FOREMAN/SUPR:	ANALYSIS BY:
DEPARTMENT: SV - SHOP		SECTION: MACHINE SHOP	REVIEWED BY:
REQUIRED AND/OR RECOMMENDED PERSONAL PROTECTIVE EQUIPMENT: SAFETY GLASSES, HAND & FOOT PROTECTION			APPROVED BY:

SEQUENCE OF BASIC JOB STEPS	POTENTIAL ACCIDENTS OR HAZARDS	RECOMMENDED SAFE JOB PROCEDURE
1. PLACE RAW STOCK BLANK IN LATHE CHUCK.	1. (SA) Sharp edges of stock blank 2. (CB) Stock and lathe 3. (SO) Lifting raw stock blank 4. (CO) Moving parts while rotating blank 5. (SB) Falling Objects	1. Wear gloves or protect against sharp edges as required when handling raw stock. 2. Watch position of hands, arms and body to keep clear of pinch points. 3. Use proper lifting procedures. Get help if blank is too large for one man. 4. Keep hands clear when rotating chuck to position part. Use jogging control or slow speed. Wear short sleeves. Do not wear gloves or rings. 5. Wear foot protection
2. FABRICATE AND/OR INSTALL TEMPLATE ON STYLUS PLATEN.	1. (F) Slips on oily surface at same or different level. Climbing for access to elevated controls of equipment.	1. Keep oil spills wiped up. Apply non-skid material to elevated steps. Use work platform engineered for job, including hand rail protection.
3. SELECT AND INSTALL TOOL CUTTER.	1. (SB) Falling objects, oily 2. (SA) Sharp tools	1. Keep parts free of oil when handling, wear foot protection. 2. Be aware of handling sharp objects.
4. SELECT MACHINE MODE	1. (CW) Electrical controls	1. Be sure machine is properly grounded and all electrical controls are in good repair.
5. MACHINE PART	1. (SB) Metal particles	1. Wear eye protection.
6. REMOVE TURNINGS AND CHIPS AS MACHINING PROGRESSES	1. (SB) Turnings and chips 2. (CO) Turnings (More)	1. Wear eye protection. Use correct tools to clean turnings from table. 2. Break turnings before they become unwieldy. Do not wear long sleeve work clothes.

Figure 7-3.—A completed JSA should become the basic guide for instructing employees. In this training guide, the abbreviations in the middle column are as follows: CB—caught between; CO—caught on; CW—contact with; F—falls; SA—struck against; SB—struck by; and SO—strain or overexertion.

120

all hazards—both those produced by the environment and those connected with the job procedure. Each step, and thus the entire job, must be made safer and more efficient. To do this, ask these questions about each step. (See Figure 7-3.)

1. Is there a danger of striking against, being struck by, or otherwise making injurious contact with an object?
2. Can a person be caught in, on, or between objects?
3. Can he slip or trip? Can he fall on the same level or to another?
4. Can he strain himself by pushing, pulling, or lifting?
5. Is the environment hazardous (toxic gas, vapor, mist, fume, or dust; heat or radiation)?

Close observation and job knowledge are required. The job observation can be repeated as often as necessary until all hazards and potential accidents have been identified. The experienced employee will probably suggest additional ideas. You should also check with others experienced with the job. Through observation and discussion you will develop a reliable list of hazards and potential accidents.

Develop solutions. The final step in a JSA is to develop a recommended safe job procedure to prevent occurrence of potential accidents. The principal solutions are:

1. Find an entirely new way to do the job.
2. Change the physical conditions that create the hazards.

3. Change the procedure of performing a task.
4. Try to reduce the necessity of doing a hazardous job, or at least, the frequency that it must be performed. This is particularly helpful in maintenance.

To find an entirely new way to do a job, determine the work goal of the job, and then analyze the various ways of reaching this goal to see which way is safest. Consider work-saving tools and equipment.

If a new way cannot be found, then ask this question about each hazard and potential accident listed, "What change in physical condition (such as change in tools, materials, equipment, or location) will eliminate the hazard or prevent the accident?" When a change is found, study it carefully to find what other benefits (such as greater production or time saving) will accrue. These benefits are additional selling points.

The third step in solving the job-hazard problem is to investigate changes in procedures. Ask of each hazard and potential accident listed, "What should a person do (or not do) to eliminate this particular hazard or prevent this potential accident?" Where appropriate, ask an additional question, "How should he do it?" In most cases, the supervisor can answer these questions from his own experience.

Answers must be specific and concrete if new procedures are to be any good. General precautions—"Be alert," "Use caution," or "Be careful"—are useless. Answers should precisely state what to do and how to do it. This recommendation, "Make certain the wrench does not slip and cause loss of balance," is only half good. It does not tell how to prevent the wrench from slipping.

The following, in contrast, is an example of a good recommended safe procedure that tells both "what" and "how": "Set wrench securely. Test its grip by exerting a slight pressure on it. Brace yourself against something immovable, or take a solid stance with feet wide apart, before exerting full pressure. This prevents loss of balance if the wrench slips."

Often a repair or service job has to be repeated frequently because a condition needs correction again and again. To reduce the necessity of such a repetitive job, ask, "What can be done to eliminate the cause of the condition that makes excessive repairs or service necessary?" If the cause cannot be eliminated, then ask, "Can anything be done to minimize the effects of the condition?"

Machine parts, for example, may wear out quickly and require frequent replacement. A study of the problem may reveal that excessive vibration is the culprit. After reducing or eliminating the vibration, the machine parts last longer and require less maintenance.

This fourth step—reducing frequency of a hazardous job—contributes to safety only in that it limits the exposure. Every effort still should be made to eliminate hazards and to prevent potential accidents through changing physical conditions or revising job procedures or both.

Finally, check or test the proposed changes by observing the job and discussing the changes with the persons who do the job. Their ideas about the hazards and proposed solutions may be of considerable value. They can judge the practicality of proposed changes and perhaps suggest improvements. Actually these discussions are more than just a way to check a JSA. They are safety contacts that promote awareness of job hazards and safe procedures.

The principal benefits that arise from job safety analysis are in these phases of a supervisor's work:

• Giving individual training in safe, efficient procedures

• Making employee safety contacts

• Instructing the new man on the job

• Preparing for planned safety observations

• Giving pre-job instructions on irregular jobs

• Reviewing job procedures after accidents occur

• Studying jobs for possible improvement in job methods

Job instruction training

The job instruction training (JIT) method described here is often called the Four-Point Method, because the instructing job is broken into four parts, each of which will be discussed in detail:

122

HOW TO GET READY TO INSTRUCT

Have a Timetable—
how much skill you expect him to have, by what date

Break Down the Job—*
list important steps, pick out the key points. (Safety is always a key point.)

Have Everything Ready—
the right equipment, materials and supplies.

Have the Workplace Properly Arranged—
just as the trainee will be expected to keep it.

*Use JSA, Job Safety Analysis breakdown to locate hazards.

JOB INSTRUCTION TRAINING (JIT)

HOW TO INSTRUCT

1. Prepare
Put trainee at ease.
Define the job and find out what he already knows about it.
Get him interested in learning job.
Place in correct position.

2. Present
Tell, show, and illustrate one IMPORTANT STEP at a time.
Stress each KEY POINT.*

3. Try Out Performance
Have him do the job—coach him.
Have him explain each key point to you as he does the job again.
Make sure he understands.
Continue until YOU know HE knows.

4. Follow-Up
Put him on his own.
Designate to whom he goes for help.
Check frequently. Encourage questions.
Taper off extra coaching and close follow-up.

*Safety is always a key point.

**SAFETY TRAINING INSTITUTE
NATIONAL SAFETY COUNCIL**

Figure 7-4.—Everyone who trains should follow the JIT format when teaching job skills.

Prepare the worker

Present the operation

Try out the performance

Follow up

The four-point method is intended to help an instructor teach a trainee to do a specific job. It aims at faster learning and better learning. When combined with a JSA, it becomes an excellent method for teaching safety along with job skills. (See Figure 7-4.)

Prepare. The instructor must first decide what is going to be taught and how the major points will be presented.

Next, the instructor must get ready the necessary tools, equipment, and materials (including necessary safety equipment). Teaching can be ruined if it does not go off smoothly.

The trainee, too, must be prepared by being put at ease. If he is going to absorb instruction, if he

123

is going to give full attention, he must not feel tense, self-conscious, or worried.

Ways to put the trainee at ease include: chatting with him, getting the student to talk about himself and his experiences, having a cup of coffee together, telling of some amusing experience, or, in short, the instructor should act glad to meet the person and anxious to get better acquainted. If the instructor cannot do this with sincerity, chances are he will not like instructing and never be very good at it.

Create interest in the job. The instructor tells the trainee what the purpose of the job is, how the job relates to other jobs in the plant or to the main business of his company, and other interesting facts about the job. He makes the person feel that learning the job will be a real accomplishment—something that he can be proud of and that will help him in the future. Nothing lets a person down as much as telling him, "There's really nothing to this job—anybody can learn it." Although this may be said in an attempt to give the trainee confidence, it really makes him feel that he should remain only as long as it takes to find a job that really amounts to something.

Instead of saying this, the instructor should tell the person that this job takes some learning but he is just the person to learn it, and that the instructor is going to stick with him to see that he does.

Relating the new job to something he has done before gives a person confidence and understanding. Careful observation and questioning will enable the instructor to know what experience the trainee already has, so that the instructor can use actual examples. The instructor can also judge better when this first step should be brought to a close. Do not spend any more time on this step than necessary.

Present. No more than five or six new points should be presented at one time. If a job consists of more than five or six parts, it can be organized into two or more units for instruction.

The instructor should both show and tell how the job is done. He begins by doing the job himself (the "regular way") for two or three times, and then, step by step, explain what he is doing. He names parts, tools, and processes carefully and accurately. He uses key points as reminders: ". . . until you can feel it strike against the gage," in a machine operation; ". . . just tight enough to turn the ratchet," . . . in using a micrometer; ". . . until it reaches this mark," in reading a gage; ". . . with your hand doubled into a fist," a safety key point in some operations; "walk your hands, don't slide them," a safety key point in handling lumber. The instructor may go through this phase of showing and telling several times slowly. He must use the same names and key points.

When the instructor shows and tells how to do a job, the trainee should observe from the approximate position of the operator. He can sit or stand by the instructor or he can look over his shoulder.

If he faces the instructor, he will see the job backward and not adjust himself to the new situation as quickly.

Try out. Now the trainee takes over and goes through the operation slowly, explaining what he is doing, using the correct names and key points taught in step 2, "present." The instructor may prompt him when necessary, but does not touch the material or equipment. To overcome temptation to do this, the instructor might keep his hands in his pockets.

If, after prompting, the trainee still is unable to go ahead on his own, the instructor goes back to step 2 and repeats the whole procedure while the trainee again observes.

As soon as the trainee can do the job without prompting, the instructor has him do it again to fix it in his mind. Then they proceed to step 4.

Follow up. The trainee does the job, without explaining, while the instructor observes and corrects if necessary. Gradually the trainee picks up speed, but not at the expense of perfect performance. If the trainee does not seem sure or makes mistakes, the instructor asks him to do the job slowly and explains each step as he goes along (step 3). If he does a good job, the instructor compliments him.

After the four-point method of job instruction training (JIT) has been completed, the instructor tells the trainee to continue turning out good work, but not yet try for speed. The instructor asks him to come back for help if he needs it, and from time to time, the instructor stops back to observe and comment on progress. He checks quality and quantity and watches closely to make sure that the trainee is not "slipping up" on necessary steps in the operation, particularly safe practices.

Neglect of safe practices is not always obvious in the finished product. It is discovered only through observation of the person at work. If the trainee is taking the necessary precautions, the instructor can well compliment him in a friendly manner, "That's the right way. You don't want to get hurt just when you're getting good on this job."

Other Methods of Instruction

The lesson plan

The method discussed so far has primarily been a method that is commonly used to teach job skills in on-the-job situations. However, there may be times when the instructor will need to know how to make a presentation or teach in a classroom situation. He may be asked to teach a part of an overall course, such as one designed to train employees to become welders or pipe fitters. There may be other occasions when it is advisable to call a group together to explain a new procedure or method. At such times, the lesson plan format will be useful.

Lesson plans are the blueprint for presenting material contained in a course outline or for presenting a single unit of instruction. In addition to standardizing train-

125

ing, lesson plans help the instructor:

- Present material in proper order
- Emphasize material in relation to its importance
- Avoid omission of essential material
- Run his classes on schedule
- Provide for trainee (student) participation
- Increase his confidence, especially if he is new

Names for the parts of a lesson plan may vary, even the order may not always be the same. The following is a good example of arrangement for a lesson plan:

1. *Title:* Must indicate clearly and concisely the subject matter to be taught.
2. *Objective:*
   a) Should state what the trainee should know or be able to do at the end of the training period.
   b) Should limit the subject matter.
   c) Should be specific.
   d) May be divided into a major and several minor objectives for each session.
3. *Training aids:* Should include such items as actual equipment or tools to be used, and also charts, slides, films, etc.
4. *Introduction:*
   a) Should give the scope of the subject.
   b) Should tell the value of the subject.
   c) Should stimulate thinking on the subject.

5. *Presentation:*
   a) Should give the plan of action.
   b) Should indicate the method of teaching to be used (lecture, demonstration, class discussion, or a combination of these).
   c) Should contain suggested directions for instructor activity ("Show flip chart," "Write key words on chalkboard").

6. *Application:* Should indicate, by example, how trainees will apply this material immediately (problems may be worked; a job may be performed; trainees may be questioned on understanding and procedures).

7. *Summary:*
   a) Should restate main points.
   b) Should tie up loose ends.
   c) Should strengthen weak spots in instruction.

8. *Test:* Tests help determine if objectives have been reached. They should be announced to the class at the beginning of the session.

9. *Assignment:* Should give references to be checked or indicate materials to be prepared for future lessons.

Each instructor should make a self-evaluation.

Programmed instruction

Programmed instruction may be used as a supplement to classroom and on-job training. Using self-contained teaching materials (including so-called "teaching machines"), programmed instruc-

126

tion permits the trainee to set an individual pace and to absorb knowledge in easy-to-take bits. The learning process is reinforced by requiring the trainee to answer questions and correct his own errors before progressing with the course.

There are a number of programmed instruction courses available in such areas as safety training, vocational training, and communications. Many of these courses use multi-media materials, such as tapes, slides, films, and teaching machines with devices.

A complete list of courses and devices is available from the National Society for Programmed Instruction, P.O. Box 137, Cardinal Station, Washington, D.C. 20017.

Independent study

Courses offered through correspondence schools are called home study courses, or independent study courses. They combine the fundamentals of good training and guidance and counseling of a qualified instructor with the convenience of studying at home or in supervised study sessions arranged by the company (often on company time).

A list of subjects taught by accredited private home study schools may be obtained from the National Home Study Council, 1601 18th Street, NW., Washington, D.C. 20009.

Extension programs of most major universities offer programs of all types through independent study and will furnish complete information upon written request.

Closed circuit TV

Closed circuit TV (CCTV) training uses television's "instant replay" techniques. The basic technique is to record the visual procedure and directions on videotapes and then play it back later on a monitor. Once the process or manual skills, along with the instructions are recorded, it can be replayed many times. It can be used to record the steps in a job safety analysis, which can then be used to train employees in safe job procedures.

CCTV is extremely flexible; portions of the tape may be shown for review purposes. It also makes for uniform training since the job is presented in exactly the same manner to each trainee.

Training may not solve all your problems, but good training will help solve many of them. The good instructor attempts to fit the training method to the problem and not use the same technique to solve all problems. In many cases, the instructor may find it advantageous to mix training methods, such as using programmed instruction along with on-the-job training to reach the highest level of learning in the shortest possible time.

Good training produces skilled and productive workers, as well as safe workers.

OSHA Training Requirements

Most employers are aware that they must comply with the standards promulgated by the Secretary of Labor under the Occupational Safety and Health Act.

However, many employers may not be familiar with those standards that require them to provide training for their employees. OSHA Bulletin, "Safety and Health Training Guidelines for General Industry" is designed to help employers and employees identify the standards that relate to training. (This bulletin is available from the OSHA area office.)

Where no "training standards" as such are identified among the occupational safety and health standards, training programs which are considered to meet the intent of the standard have been identified in Bulletin 2082.

An example is Part 1910.151 (Medical Services and First Aid of Subpart K). The Occupational Safety and Health Administration (OSHA) has agreed that completion of the basic American National Red Cross course in first aid will be considered as having met the requirements that "a person or persons shall be adequately trained to render first aid . . . ." However, Regional Administrators may determine whether other courses meet the requirement.

Although OSHA does not certify or approve training programs, OSHA personnel conducting compliance visits to workplace establishments will look for evidence that an employer has provided the training required.

In the absence of definitive "training standards," the employer should be able to produce records indicating that his employees have received training in the areas identified in the standards.

In addition, the employer should be able to show the compliance officer that training has been given to the employee based on an analysis of the tasks performed by the employee. Such task analysis should identify, as a minimum, the actual and potential hazards the employee would encounter on the job, and the equipment and practices the employee should use to minimize the risk of injuring himself or another employee.

The employer also should be able to show that the training gives priority to the types of conditions and practices most likely to result in injury and illness.

Promotion, Motivation, and Employee Involvement

CHAPTER 8

How To Promote Safe Workmanship

Objectives and benefits

Maintaining interest in safe workmanship is necessary even if the work place is safe, even if work procedures are designed with safety in mind, and even if supervisors train their men thoroughly and enforce safe work procedures. Why is it still necessary to maintain interest? Because even with optimum work conditions, accident prevention depends upon a continuing desire of people to want to work safely.

There must be a safety awareness that leads to a positive attitude that builds a potential to act safely everywhere—all the time. Each employee must be stimulated to think beyond his immediate work situation in order to respond safely in questionable situations when he is "on his own."

Just as products and services require imaginative sales promotion, safety requires skillful promotion. A well-planned safety motivation program should:

• Help develop safe work habits and attitudes

• Focus attention on the causes of accidents and help identify and eliminate them

• Supplement (but not substitute for) safety training

• Let employees participate in suggesting safety improvements and safer job procedures

• Provide communication channel between employees and supervisors

• Improve employee, customer, and public relations

Safety committees

One of the best ways to create and maintain interest is to get employees personally involved in the safety program. Safety committees are one way to stimulate interest in accident prevention. They can provide a place for discussion of accident problems and preventive measures and they can help a manager evaluate safety suggestions of employees. They promote intelligent understanding of problems.

Safety committee meetings can be scheduled so they do not interfere with operations or services. Membership should be limited

129

and periodically changed. Members of the committee can be given assignments such as inspections, observing work practices, and accident investigation. A record of their recommendations should be kept.

The success of a safety committee depends on the way the boss handles it. The meetings should be business-like and not turn into a gripe session.

Safety meetings

Safety meetings should be used for communicating with or training employees and/or supervisors. Meetings should not be held unless there is a need, such as the following:

• Supervisors and management meet to discuss and adopt policies, initiate programs, or plan special safety activities.

• All employees meet for a special purpose, such as training or being informed of something.

• Departments or special groups meet to discuss problems or introduce new equipment or procedures.

Here are some basic facts about meetings that will help improve them.

• People can absorb only a few ideas at a time.

• Meeting facilities should be comfortable and away from distracting noises.

• Meetings should be carefully planned. Review material to be presented. Check audiovisuals ahead of time to make sure they work properly.

• Keep meetings to 30 minutes or less.

"Stand up meetings" held at the beginning of a work shift or before tackling a new job can provide a shot in the arm for safety.

Audiovisuals, exhibits, and demonstrations make a meeting more interesting, especially if they are presonalized to the actual job situation.

Contests and competitions

Contests and competitions create interest in safety—but they are not a substitute for a basic safety program. Contests are based on accident experience and are operated for a stated period. An incentive or prize for measurable improvement can be offered.

Be sure to indicate a period of time that the contest covers—such as July 1 to December 31. Or, instead of a six-month contest, have a series of monthly contests culminating in a grand winner at year-end.

Because the purpose of contests is to stimulate interest, there is virtually no limit to the number of ways that good showmanship and imagination can be applied.

Remember that *good* performance should be awarded; but at the same time, a bad safety performance should not be ridiculed.

Suggestion systems

A well-organized suggestion system encourages employees to contribute their ideas to improve work methods or reduce work hazards or both. It stimulates their thinking about solving production and safety problems. If an idea is adopted, it should be

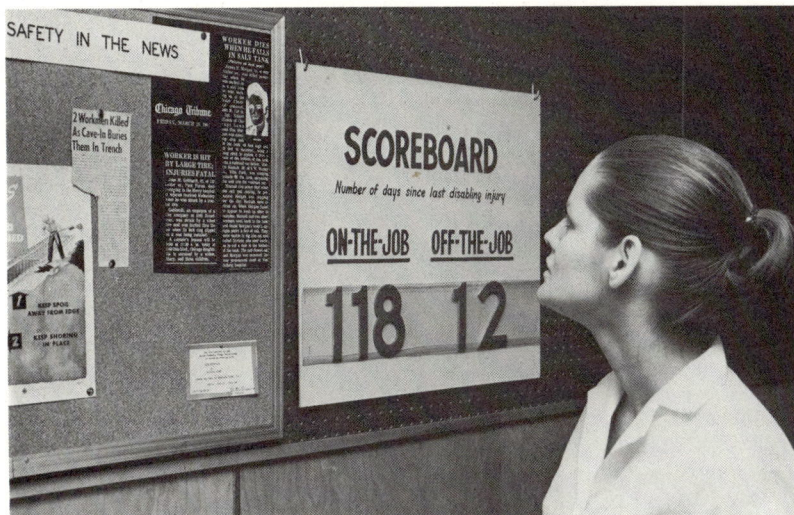

Figure 8-1.—Contests can be between departments or between plants. Signs with changeable numbers keep employees informed of safety performance. Safety bulletin board (*left*) can include posters, relevant news items, and notices.

awarded according to the money it saves or the seriousness of the accident it prevents.

The successful suggestion system has a clearly stated and easily understood operating plan that is scrupulously observed. Employees must know that the plan is fair, impartial, and potentially profitable for themselves.

Posters and displays

Because safety posters and signs are the most visible evidence of a company's interest in accident prevention, some firms have mistakenly assumed that posters alone will do the safety job. Actually, hit or miss use of posters where no other safety work is done is more likely to have a negative influence because employees do not feel the company is sincere.

When properly used, posters can influence safety attitudes. They can be compared to commercial advertising that is used to influence buying behavior. Safety posters and eye-catching safety displays alert people to safe practices.

Posters should have high visibility yet be located so they will not interfere with traffic. They should be centered at eye level in well-lighted areas. National Safety Council posters are available in two sizes—17 by 23 inches and 8½ by 11 inches. Poster mounting areas should be planned accordingly. Posters may be displayed alone or with other notices, if they are kept free of clutter.

Posters and displays should be changed regularly. Choose subject matter that is related to problems or bad practices, campaigns, or holidays. NSC's biennial Catalog and Poster Directory contains

131

a wide selection of posters covering a variety of hazards and motivational subjects.

Other means of publicizing safety messages include displays of personal protective equipment for employees to learn about and displays of safety equipment that has prevented injury to their wearers. Special demonstrations of how safety and fire prevention equipment work help generate confidence and use by employees.

Safety literature

The National Safety Council, trade associations, insurance companies, and professional organizations publish a wide variety of booklets and leaflets that are authoratative, attractive, and informative. These can be distributed to motivate employees to work safely.

How To Motivate Workers To Be Safe

The human factor operates at all levels and is perhaps the most potent factor for success or failure of a safety program. It makes a difference whether the president of a company approves or drags his feet, whether the safety professional works hard or coasts along, whether the supervisor emphasizes safety or subordinates it to production, whether the janitor cleans well or does only the minimum—attitudes are important to safety in the company. Safety can be achieved only by working through all these people.

To achieve some mark of success in dealing with people, individuals can best be considered within the framework established in this discussion. Each person is an individual, and to some degree, different from every other one. The differences are for the most part obvious, but there are subtle ones too, which must be recognized, if only to the extent that their existence is acknowledged.

Despite great differences in people, reasons for their activities are common to all. Many needs are the same, particularly at the biological level and to a large degree at the psychosocial level. It is upon these needs that the safety professional and others in positions of leadership in industry can capitalize to most effectively promote safety.

People become frustrated when their goals cannot be achieved. There are many ways of reacting to such frustrations, but the emotional reactions are of great concern to the safety professional. These reactions, as well as the attitudes formed during the reactions, can be highly disruptive to safety precautions and procedures.

Training programs are established to teach safe work methods. These learning situations, if they are to operate efficiently, call for knowledge of the basic principles of learning that cause people to learn and act as they do. These principles should be used to make new learning more efficient.

Safety devices, safer machines, safe work layout, and many other aids and measures all are a part of the total program. The human factor is one more aspect of the whole system.

People, machines, and materials are still the three components of industry that can contribute to safety.

Machines and materials can be controlled but the human factor must be guided in the interests of accident prevention.

Understanding people

To work effectively with people, one must first understand them. To gain such understanding, one must examine those individual personality factors that set each individual apart from all others. The basic question is, "What factors common to people can be utilized to make a safety effort effective?"

There are at least five major psychological factors of people which affect their participation in any accident prevention effort and determine the results obtained:

• Every individual is different.

• People have different motivations.

• Each individual has a different emotional makeup and reaction level.

• Peoples' attitudes vary and are changeable.

• The learning processes of individuals are different.

Each of these factors may be involved in the cause of any one accident, and each of the factors has many variations. Since small businesses usually cannot afford to experiment in the psychological area, certain basic principles must be accepted as a general guideline for management decisions.

Job motivators and dissatisfiers

Studies show that workers are motivated or dissatisfied on the job depending upon the events and circumstances involving the following elements:

Company policy and administration

Recognition

Competence of supervision

Advancement and possibility of growth

Responsibility

Pay

The work itself

Attitude (example) of supervisors

Peer relations

Working conditions

Status

Job security

People are motivated to act in order to satisfy their physiological needs and their social needs. The physiological needs are primarily those of food and shelter. Most theories say that once these basic physiological needs are satisfied, the social needs become more powerful. Social needs such as (a) knowing work has value, (b) appreciation, (c) security, and (d) feeling of belonging are job motivators. Pride of accomplishment is also a strong motivator.

To motivate the worker to carry out his duties safely, safety must become an integral part of his job. The worker must be made to realize that advancements and recognition come to the safe and productive worker. Achievement must not be measured in product or service, but rather in terms of both product or service with safe-

133

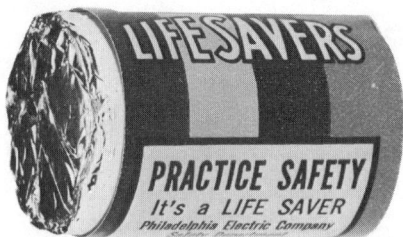

Figure 8-2.—It's name makes this well-known confection a "natural" for conveying a safety message. Paper cups, pencils, notebooks, and other items have also been used with good effect.

pervision, company policy, chance to advance, working conditions, pay, responsibility, recognition status, job security and the job itself there are many variables involved in his reaction.

The attitudes of workers may be affected by the example of the employer and training and indoctrination of the employee. They can also be affected by the supervisor's ability to work with people and to show them that their work has value and their work is appreciated.

ty. If safety is presented as an integral part of the job, then safety will become part of the worker's and supervisor's motivational system.

Attitudes and emotions

Attitudes are so closely allied to emotions that some people consider them as pure emotional responses.

Strictly defined, an attitude is a predisposition on the part of a person to respond in a given way. Naturally, in safety, it is hoped the natural response to a hazardous situation will be the safe response. Because attitudes play such an important role in everyday relationships, consideration should be given to their development, their effects on individuals, and particularly on what can be done to change them—if they are wrong.

A worker's attitude is based on years of past experiences plus his present environment. The past experience may have been good or bad, painful or happy. As he tries to relate them to his present su-

Morale and team spirit

People want to belong to their work group and they want to feel secure in their jobs. Thorough training in their assignments helps give employees a feeling of security, and will provide for more safe and efficient production. Injuries, work spoilage, and misused or broken equipment are often the result of poor morale.

Managers and supervisors should watch for opportunities to talk to their employees and to create good job interest. Using words such as "we," "our section," and "our safety record" encourages group interest and a team spirit.

The group feeling or team spirit promotes good performance and safe practices. Once group spirit develops, members discipline each other. Often the group standards are higher than the standards of many of the individuals by themselves.

Incentive awards

Incentive awards for safety achievements can vary as much as

134

Courtesy National Dairy Products Corporation.

Figure 8-3.—Presentation of a National Safety Council award provides a good occasion to interest and encourage employees to try for even greater safety achievement.

a person's imagination and ingenuity. The originality or cleverness of an award, or in the manner of its presentation is a very important factor. (See Figure 8-2.)

A free cup of coffee to everyone on achieving a number of injury-free hours would probably draw more favorable response than a fancy plaque to the supervisor. The drawing of a small cash prize or a grab bag prize would attract more interest than a routine presentation of the same award. An award to an employee's wife for completing a home safety checklist, for identifying a safety slogan, or for contributing to the company paper would create more interest than the same award given on the job.

An award serves several purposes. It is an inducement, a builder of good will, a continuing reminder, and a basis for publicity. To serve these purposes, however, an award must be meaningful.

The value of awards lies in their appeal to basic interest factors such as pride, need for recognition, urge to compete, and desire for personal gain.

This discussion of motivation is of necessity brief. The subject of motivation and the relationship of human relations to safety are covered in depth in a chapter of NSC's *Accident Prevention Manual for Industrial Operations* and in the *Supervisors Guide to Human Relations.* How to promote safe workmanship through improved communications is the subject of *Communication for the Safety Professional.*

135

Safeguarding Machines, Tools, and Equipment

CHAPTER 9

Machines and equipment are the lifeblood of industry, but they do not need the life blood of people to make them work. What they need is human intelligence and ingenuity to make them useful to mankind. The application of the principles of safeguarding, inspection, and maintenance given here will do much to eliminate accidents and keep production running smoothly.

Need for Safeguarding Machinery

Mechanical equipment and production machinery must be kept running efficiently if maximum production benefits are to be obtained. When equipment is shut down for any unscheduled reason, production capacity and cost of operation are directly involved. When an employee is injured, production capacity and cost of operation are again affected.

Injuries resulting from machines and equipment operations can be quite serious and quite costly. An injury can take an operator away from his job for a long time—sometimes permanently. This person must be replaced if the operation is to keep going.

Finding this person takes time. Hiring takes time. Training takes time. And time costs money.

How does safeguarding fit into this picture? No matter how much training a person may get and no matter how much experience he may have, he can still be injured by an unguarded machine. Safeguarding reduces the possibility of accidents involving "human error," mechanical failure, poor design, and electrical failures. Owners and managers should have a good understanding of safeguarding principles and rules and regulations that apply to their operations.

Principles of Safeguarding

Moving parts of machinery and equipment can trap fingers, hands, feet, or other parts of the body. Bits and pieces of materials in process can be thrown. Gases, vapor, and/or heat can be emitted. In addition there can be a hazard from sharp edges on in-process materials, and a hazard from just moving the materials. The energy source for the machinery and equipment can, in itself, be hazardous; even a nonfatal shock could startle the operator and

Courtesy Service Pipe Line Co., Tulsa, Oklahoma.

Figure 9-1.—An expanded metal screen encloses the V-belt drive of this injection pumping unit and its 200-hp explosion proof electric motor.

cause him to put part of his body in jeopardy.

For efficient and safe operation, hazards should be eliminated or minimized by engineering design. If not, hazards must be safeguarded. Employees must be educated and trained to make sure control measures are effective.

Applied engineering

As discussed in Chapter 7 under Job Safety Analysis, hazards can frequently be eliminated by analyzing each step of the manufacturing process and asking, "Is this step really necessary?" If it is, then the next question should be, "Can it be done more efficiently and safely?" Can a different machine, process, or material be substituted? Can operational steps be combined? Can automation or other mechanical assistance be applied to eliminate the need for human involvement near a hazard point? If automation cannot be applied, can other mechanical changes be made—such as relocating a lubrication fitting from an internal location to an external one?

Safeguarding

Where hazard points cannot be eliminated, safeguarding must be considered. Many times, all that is required is placing a barrier in front of the hazard point. Today, machine and equipment manufacturers are doing this, especially for power-transmission components. Usually a safeguard, built into a machine or equipment, is

137

more effective and stronger than one added on later in the shop. Such safeguarding should be specified whenever new equipment is purchased.

Older equipment may lack guarding or it may be inadequate or damaged. Often it has been taken off and not replaced. Whatever the reason, these guards must be replaced and used.

Safeguarding the point of operation is sometimes the most difficult guarding job of all and may require much engineering know-how. The point-of-operation is frequently the point of greatest hazard because this is the place where man and machine are most closely involved. A fixed barrier guard is the most effective because if a worker cannot get into the point of operation, he cannot be exposed to the hazard. Even machines that are operated automatically (that require only servicing of stock or monitoring) should have barrier guards to prevent operators, as well as passers-by from inadvertently entering the hazard point.

Some jobs preclude the use of fixed barrier guards because of the size or configuration of the part or because of frequent job changes. This does not eliminate the need for safeguarding. Most alternate safeguarding allows access to the point of operation between cycles for loading and unloading. Extra precaution is required to protect against accidental operation from whatever cause.

It is a cardinal rule that safeguarding one hazard should not create an additional hazard. The created hazard is frequently hidden, and if it causes an accident, it is because people are part of the system. People are limited in their capabilities—some are forgetful and others are very attentive. If these individual differences are not allowed for when designing a system, they can and will lead to accidents. A forgetful person may inadvertently enter a hazard area because he "forgot about its being there." An attentive person may inadvertently enter a hazard area because he was concentrating on something else.

These are only two examples of many other mental, emotional, psychological, and physical limitations and capabilities that must be considered when designing a safeguarding system. This approach to safeguarding is called "human factors engineering." It is vital for successful accident-free operations.

Some safeguarding devices that will provide effective safeguarding are listed. These can be used alone, but more frequently they are used in combination. Some of these are: adjustable barriers, presence-sensing devices, pull-outs or restraints, and hand-feeding tools.

There are many differences between various guards and safeguards; refer to Table 9-A. *Guards* permanently prohibit entry into hazard points during normal operating procedures. They should only be removed for authorized service or maintenance and then with special tools and under lockout procedures. They must be replaced after the service or maintenance is com-

TABLE 9-A. POINT-OF-OPERATION PROTECTION

Type of Guarding Method	Action of Guard	Advantages	Limitations	Typical Machines on Which Used
Enclosures or Barriers				
Complete, simple fixed enclosure	Barrier or enclosure which admits the stock but which will not admit hands into danger zone because of feed opening size, remote location, or unusual shape.	Provides complete enclosure if kept in place. Both hands free. Generally permits increased production. Easy to install. Ideal for blanking on power presses. Can be combined with automatic or semiautomatic feeds.	Limited to specific operations. May require special tools to remove jammed stock. May interfere with visibility.	Bread slicers Embossing presses Meat grinders Metal square shears Nip points of inrunning rubber, paper, textile, and other rolls Paper corner cutters Power presses
Warning enclosures (usually adjustable to stock being fed)	Barrier or enclosure admits the operator's hand but warns him before danger zone is reached.	Makes "hard to guard" machines safer. Generally does not interfere with production. Easy to install. Admits varying sizes of stock.	Hands may enter danger zone—enclosure not complete at all times. Danger of operator not using guard. Often requires frequent adjustment and careful maintenance.	Band saws Circular saws Cloth cutters Dough brakes Ice crushers Jointers Leather strippers Rock crushers Wood shapers
Barrier with electric contact or mechanical stop activating mechanical or electric brake	Barrier quickly stops machine or prevents application of injurious pressure when any part of operator's body contacts it or approaches danger zone.	Makes "hard to guard" machines safer. Does not interfere with production.	Requires careful adjustment and maintenance. Possibility of minor injury before guard operates. Operator can make guard inoperative.	Dough brakes Flat roll ironers Paper box corner stayers Paper box enders Power presses Rubber and paper calenders Rubber mills
Enclosure with electrical or mechanical interlock	Enclosure or barrier shuts off or disengages power and prevents starting of machine when guard is open; prevents opening of the guard while machine is under power or coasting. (Interlocks should not prevent manual operation or "inching" by remote control.)	Does not interfere with production. Hands are free; operation of guard is automatic. Provides complete and positive enclosure.	Requires careful adjustment and maintenance. Operator may be able to make guard inoperative. Does not protect in event of mechanical repeat.	Dough brakes and mixers Foundry tumblers Laundry extractors, driers, and tumblers Power presses Tanning drums Textile pickers, cards

Type of Guard-ing Method	Action of Guard	Advantages	Limitations	Typical Machines on Which Used
	*Automatic or Semiautomatic Feed*			
Nonmanual or partly manual loading of feed mechanism, with point of operation enclosed	Stock fed by chutes, hoppers, conveyors, movable dies, dial feed, rolls, etc. Enclosure will not admit any part of body.	Generally increases production. Operator cannot place hands in danger zone.	Excessive installation cost for short run. Requires skilled maintenance. Not adaptable to variations in stock.	Baking and candy machines Circular saws Power presses Textile pickers Wood planers Wood shapers
	Hand Removal Devices			
Hand restraints	A fixed bar and cord or strap with hand attachments which, when worn and adjusted, do not permit an operator to reach into the point of operation.	Operator cannot place hands in danger zone. Permits maximum hand feeding; can be used on higher-speed machines. No obstruction to feeding a variety of stock. Easy to install.	Requires frequent inspection, maintenance, and adjustment to each operator. Limits movement of operator. May obstruct work space around operator. Does not permit blanking from hand-fed strip stock.	Embossing presses Power presses
Hand pull-away device	A cable-operated attachment on slide, connected to the operator's hands or arms to pull the hands back only if they remain in the danger zone; otherwise it does not interfere with normal operation.	Acts even in event of repeat. Permits maximum hand feeding; can be used on higher speed machines. No obstruction to feeding a variety of stock. Easy to install.	Requires unusually good maintenance and adjustment to each operator. Frequent inspection necessary. Limits movement of operator. May obstruct work space around operator. Does not permit blanking from hand-fed strip stock.	Embossing presses Power presses
	Two-Hand Trip			
Electric	Simultaneous pressure of two hands on switch buttons in series actuates machine.	Can be adapted to multiple operation. Operator's hands away from danger zone.	Operator may try to reach into danger zone after tripping machine. Does not protect against mechanical repeat unless blocks or stops are used. Some trips can be rendered unsafe by holding with the arm, blocking or tying down one control, thereby permitting one-hand operation. Not used for some blanking operations.	Dough mixers Embossing presses Paper cutters Pressing machines Power presses Washing tumblers
Mechanical	Simultaneous pressure of two hands on air control valves, mechanical levers, controls interlocked with foot control, or the removal of solid blocks or stops permits normal operation of machine.	No obstruction to hand feeding. Does not require adjustment. Can be equipped with continuous pressure remote controls to permit "inching." Generally easy to install.		

140

Type of Guarding Method	Action of Guard	Advantages	Limitations	Typical Machines on Which Used
		Miscellaneous		
Limited slide travel	Slide travel limited to ¼ in. or less; fingers cannot enter between pressure points.	Provides positive protection. Requires no maintenance or adjustment.	Small opening limits size of stock.	Foot power (kick) presses Power presses
Electric eye	Electric eye beam and brake quickly stop machine or prevent its starting if the hands are in the danger zone.	Does not interfere with normal feeding or production. No obstruction on machine or around operator.	Expensive to install. Does not protect against mechanical repeat. Generally limited to use on slow speed machines with friction clutches or other means to stop the machine during the operating cycle. Can be circumvented	Embossing presses Power presses Rubber mills Squaring shears Press brakes
Special tools or handles on dies	Long-handled tongs, vacuum lifters, or hand die holders which avoid need for operator's putting his hand in the danger zone.	Inexpensive and adaptable to different types of stock. Sometimes increases protection of other guards.	Operator must keep his hands out of danger zone. Requires unusually good employee training and close supervision.	Dough brakes Leather die cutters Power presses Forging hammers
Special jigs or feeding devices	Hand-operated feeding devices of metal or wood which keep the operator's hands at a safe distance from the danger zone.	May speed production as well as safeguard machines. Generally economical for long jobs.	Machine itself not guarded; safe operation depends upon correct use of device. Requires good employee training, close supervision. Suitable for limited types of work.	Circular saws Dough brakes Jointers Meat grinders Paper cutters Power presses Drill presses

pleted. *Safeguards,* on the other hand, restrict entry to hazard points by interacting with the operator during the operation. They depend on the reliability of all components of the system. Thus, a failure of either (or both) the human or the machine components can produce an accident if it occurs during entry into the point of operation.

Most safeguards are designed to be "fail safe" and many are successful up to a point. The problem lies in the human control required to make the system work. If the safeguard is not used, not adjusted, not maintained, or bypassed (deliberately or not), it cannot do its job and sooner or later an accident will occur.

No guard or safeguard is fool-

141

proof. It can be defeated if desired. A permanent barrier guard once removed for maintenance can be left off. This condition frequently is more dangerous than having a known open hazard because the operator assumes that the guard is still there.

Education and training

From the foregoing discussion, it is evident that even the best safeguarding program can fail because of human error—haste, inattention, and ignorance being the most common. In the opinion of many, ignorance may be the greatest source of error. Statements such as, "I didn't know," "I didn't think it mattered," or "No one told me," are repeated over and over again by people who have been injured.

Fortunately, ignorance can be replaced by understanding and knowledge if education and training are made equal partners in the safeguarding program. The elements of job safety analysis and job training were discussed in Chapter 7, Training.

OSHA Regulations That Apply to Machine Safeguarding

Neglect of any of the principles of safeguarding can only result in accidents that waste valuable human and material resources. In addition, these accidents may provoke an OSHA inspection. Under the OSHAct, employers are required to maintain a safe and healthful workplace; serious accidents are evidence to the contrary. In general, OSHA safety and health compliance officers check to make sure that the minimum standards are complied with.

OSHA regulations (Title 29, *Code of Federal Regulations*, Part 1910, Safety and Health Standards for General Industry) contain three major subparts that deal primarily with machine guarding—Subpart O, Machinery and Machine Guarding; Subpart P, Hand and Portable Powered Tools and Other Hand-Held Equipment; and Subpart R, Special Industries.

The guarding requirements differ only in that some machines have specific requirements that must be met. All machines are covered under the "general duty" clause:

One or more methods of machine guarding shall be provided to protect the operator and other employees in the machine area from hazards such as those created by nip points, rotating parts, flying chips and sparks. Examples of guarding methods are— barrier guards, two-hand tripping devices, electronic safety devices, etc. (§1910.212(a)(1).)

As can be seen, the intent of the regulation is that the most suitable method of protecting the operator or other employees must be used, and it applies both to point of operation and to power transmission hazards, except for maintenance and service. This requirement is easy to meet if the guards are equipped with interlocks that require the guards to be in place in order for the machine to be operated. If interlocks are not feasible, a lockout/tagout system must be used to assure the safety of operating and maintenance personnel.

142

Inspection and maintenance records for machines and equipment should be kept and made available to the OSHA compliance safety and health officer on request. Even though OSHA requires only these records in some instances, they can *ask* for them in any instance and it would be a good idea to have them. The same holds true for training records. If the compliance officer asks if you have a training program, it is not enough to reply, "Oh, yes, we have one," because he will require written proof of your program and its use.

In general, these records are simple to set up and be concise enough to keep on a chart (or keysort file or computer, if available). All that is needed is a list of various maintenance jobs for each machine or piece of equipment and the interval of their performance; in a separate column, indicate the date of actual performance of the maintenance.

As for training, a list of operators and a description of their training, with accompanying dates, should be sufficient. Be sure to keep the records up to date.

Keeping records of maintenance and training should be of more value to you than to the OSHA compliance officer. The records will pinpoint strengths and weaknesses in the mechanical and human components of your operations.

The compliance officer may ask to see the plant general work rules. It is best to have them type-written (or printed) and either posted conspicuously throughout

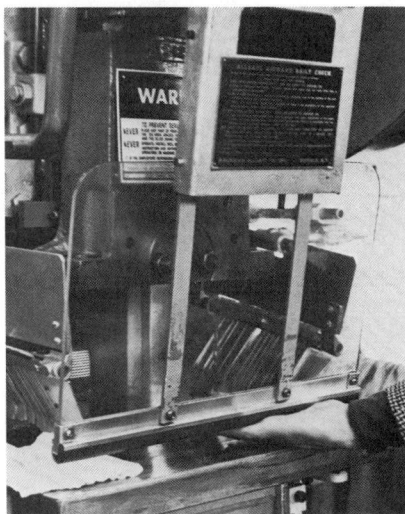

Figure 9-2.—This automatic protection device will stop the ram if the operator's fingers are near the danger zone when the press is activated. Tough, transparent plastic allows an unobstructed view of the work. An interlock (not shown) prohibits press operation if, for some reason, the device is not operating properly.

the plant or incorporated in an employee handbook.

With respect to machine and equipment operation he will be looking for rules similar to the following:

• No machine shall be operated by personnel other than the regularly assigned operator, without supervisory knowledge and prior approval.

• No machine shall be started unless the guards are in place and in good condition. Defective or missing guards shall be reported to the supervisor immediately.

• No guard shall be adjusted or removed for any reason by anyone, unless (*a*) specific permis-

143

sion is given by the supervisor, (*b*) the person concerned is specifically trained, and (*c*) machine adjustment is considered a normal part of his job.

• Whenever safeguards or devices are removed in order to repair, adjust, or service equipment (such as lubrication and maintenance), the power for the equipment must be turned off and the main switch locked and tagged.

• Employees are not permitted to work on or around mechanical equipment while wearing neckties, loose clothing, watches, rings, or other jewelry.

• Personal protective equipment will be worn by personnel on jobs requiring it.

These are the main requirements. For special machines or conditions, the requirements are more exacting. All employers should, therefore, have a copy of the requirements of the Occupational Safety and Health Administration and familiarize themselves with all portions that apply to their operations. Understanding these standards can frequently be helped by referring to the complete American National Standard from which they were drawn. The complete standard has much explanatory material and illustrations that are not included in the OSHA version. The cross references for specific requirements of Subpart O, Machinery and Machine Guarding, are as follows:

§1910.213. ANSI O1.1-1954 (R 1971), *Safety Code for Woodworking Machinery.*

§1910.214. ANSI O1.1-1954 (R 1961), *Safety Code for Woodworking Machinery.*

§1910.215. ANSI B7.1-1970, *Safety Codes for Abrasive Wheels.*

§1910.216. ANSI B28.1-1967, *Safety Code for Mills Calenders in the Rubber and Plastic Industries.*

§1910.217. ANSI B11.1-1971, *Safety Standard for Construction Care and Use of Mechanical Power Presses.*

§1910.218. ANSI B24.1-1971, *Safety Standard for Forging.*

§1910.219. ANSI B15.1-1953 (R 1958), *Safety Code for Mechanical Power Transmission–apparatus.*

OSHA compliance officers also use additional ANSI standards for guidance when making inspections. Contact the American National Standards Institute for latest standards that apply to your specific operations.

Inspecting for Safety and OSHA

Machine and equipment inspections are vital to an accident prevention program and must be regularly scheduled. Even though a company may have been making the same product or rendering the same service for many years, working conditions are seldom static. Sometimes change is so gradual that it is not noticed until an accident occurs. It is not the purpose of this discussion to repeat what has been said in Chapter 4, Safety Inspections, but rather to pinpoint what an inspector would search for were he to check the machines and equipment.

144

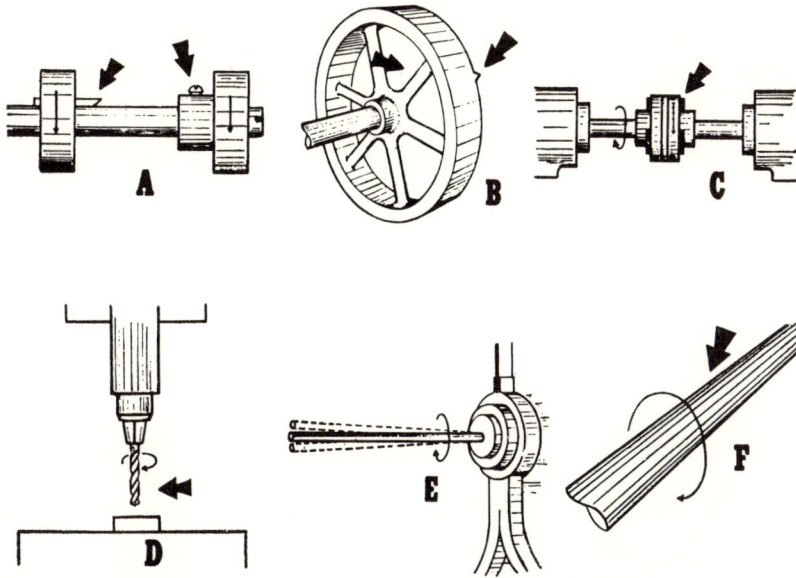

Figure 9-3.—Rotating mechanisms can seize and wind up loose clothing, belts, hair, and the like. They should, therefore, be guarded. Left to right, they are: (A) projecting key and setscrew, (B) spokes and burrs, (C) coupling bolts, (D) bit and chuck, (E) turning bar stock, and (F) rotating shaft.

• Is it possible for a person to come into direct contact with a moving machine part in normal production or maintenance operations?

• Are rotating or moving screws, keys, bolt heads, burrs, or other projections so exposed as to snag a worker's clothing? (Figure 9-3.)

• Is it possible to be drawn into the in-running nip point between two moving parts, such as a belt and sheave, chain and sprocket, pressure rolls, rack and gear, or gear train? (Figure 9-4.)

• Do machines or equipment have reciprocating movement or any movement where workers can be caught on or between a moving part and a fixed object?

• Is it possible for a worker's hands or arms to make contact with moving parts at the point of operation where milling, shaping, punching, shearing, bending, grinding, or other work is being done?

• Is it possible for material (including chips or dust) to be kicked back or ejected from the point of operation and injure someone nearby?

• Are machine controls safeguarded to prevent unintended or inadvertant operation? Are they located to provide immediate power cutoff in case of emergency?

• Do machines vibrate, move, or walk while in operation?

145

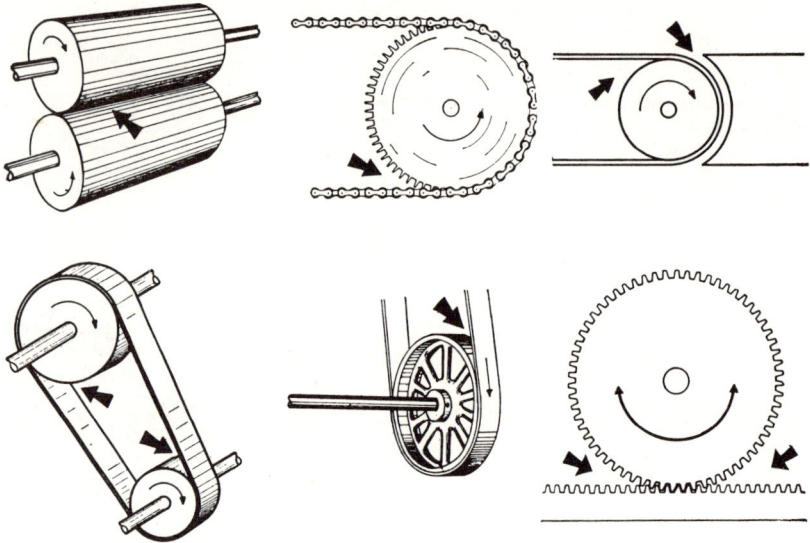

Figure 9-4.—Typical inrunning nip points. Protection against such hazards is required.

• Is it possible for parts to become loose during operation and fall on operators?

• Are guards positioned to correspond with the permissible openings between the bottom edge of the guard and the feed table? (See Figure 9-6.)

• Is it possible for the operator to bypass the guard so as it make it ineffective?

• Do machines, equipment, and appurtenances appear to have regular maintenance?

The answers to these questions are significant and may be the basis for an OSHA citation in addition to an accident. Both can be avoided by regular inspection and maintenance.

OSHA inspectors are concerned with other conditions in machine areas because machines do not stand in isolation, but are part of the total work environment.

• Are machines placed so that operators have sufficient room to work with no exposure to aisle traffic?

• Is there sufficient room for maintenance and repair?

• Is there sufficient room to accomodate incoming and finished work as well as scrap that may be generated?

• Are the materials handling methods adequate for the work in process and the tooling associated with it?

• If tools, jigs, or other work fixtures are required, are they stored conveniently, but where they will not interfere with the work?

• Is the work area well illuminat-

146

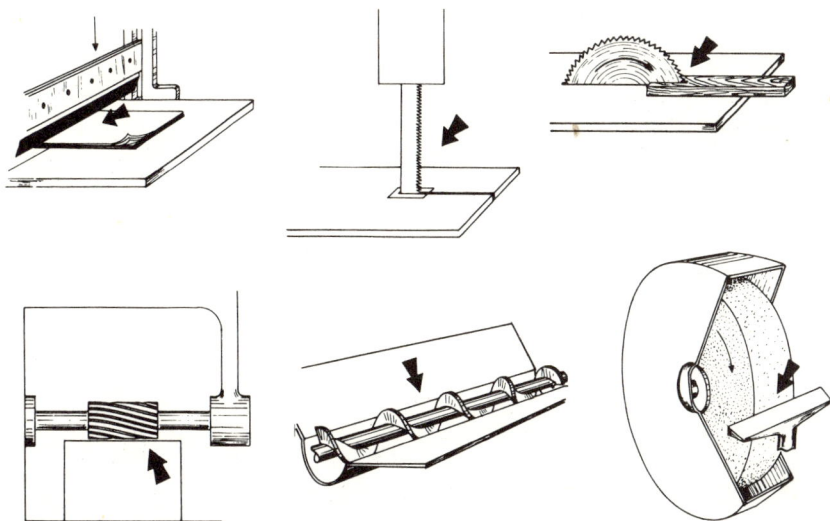

Figure 9-5.—Common cutting and shearing mechanisms. Protection against all variations of such hazards should be provided.

ed, with additional point-of-operation lighting where necessary?

• Is ventilation adequate, particularly for those operations that create dusts, mists, vapors, fumes, and/or gases?

• Is the operator using personal protective equipment if the work process indicates a need for it?

• Is housekeeping satisfactory, with no debris or tripping hazards or spills on the floor?

All of these questions will be of most value if they are incorporated into an inspection form which can be filled out at regular intervals. Even though a question may not at first seem to apply to a specific operation, on closer scrutiny, it may be found to apply. Using a checklist is a good way to make sure nothing is overlooked.

Maintenance procedures

Periodic maintenance is required to keep machines and equipment functioning properly. Maintenance or service should never be performed while a machine is operating or the power is on; this only invites accidents. Unfortunately, there are companies that allow work to be done under these conditions as a matter of course. For example, oiling and cleaning are frequently done while a machine is operating; sharpening dies is frequently done while the power is on or off without using safety blocks.

Most machines and equipment today are operated by electricity, but some of this energy is converted into other energy forms which can booby trap a maintenance man even if he does shut off the electricity and lock it out.

147

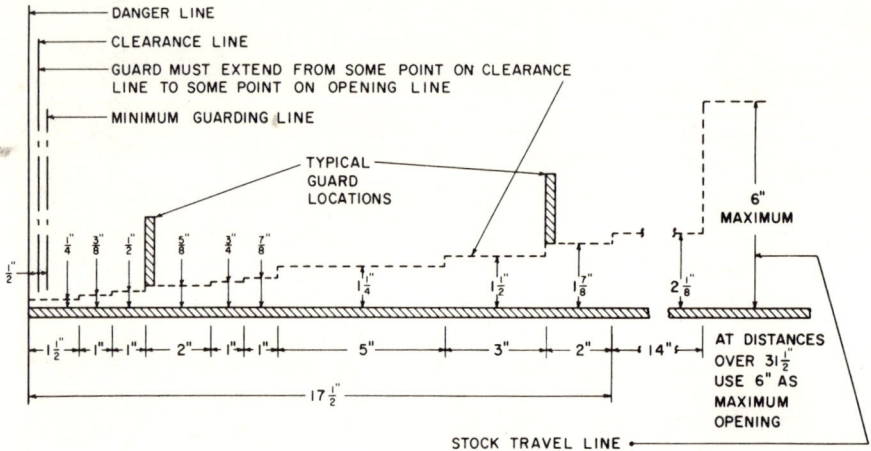

From ANSI B11.1, "Safety Requirements for the Construction, Care, and Use of Mechanical Power Presses."

Figure 9-6.—Point of operation guard locations. Barriers placed to touch dashed line will wedge hand or forearm. The danger line is the point of operation. The clearance line marks the distance required to prevent contact between the guard and the moving-parts. The minimum guarding line is 1/2 in. from the danger line.

Machines with their power off and at rest still contain stored (potential) energy in various forms—pneumatic and hydraulic systems may be under pressure; springs may be compressed or under tension; chemicals can react; and gravity remains a constant force that affects all objects.

Good maintenance procedures require that all energy sources in a machine, equipment, or process be recognized and procedures instituted to bring them to a Zero Energy State (ZES) before starting maintenance work in any given area. In most maintenance work, this means completely neutralizing or isolating all energy sources from the area being worked on, unless doing so would create additional and greater hazards.

Zero energy state (ZES) procedures assure the safety of all operating, service, and maintenance personnel by making it impossible to activate a machine. ZES procedures, however, are effective only if management trains employees to follow the procedures and checks to see that they actually do so. A typical ZES routine is described next.

Assign each person a lock(s), key(s), and lockout device(s). No two locks should have the same key. The employee's initials or clock number should be stamped on his lock(s) or on a metal tag attached to the lock(s). A lockout device is shown in Figure 9-7.

The procedure to deenergize equipment is as follows:

1. Notify the operator that repair work is to be done on his

148

machine.

2. Turn electrical power off and place lockout device through the holes in the power handle and flanges on the box.

3. Place individual padlock on the lockout device. Others who are working on the same equipment must add their individual locks to the same device.

4. Place MAN AT WORK tag at the control(s).

5. Make sure that all moving parts are at rest.

6. Check for pneumatic, hydraulic, or other fluid lines in the operation. If they affect the area under maintenance, bleed, drain, or purge them to eliminate pressure, contents, or both.

7. Valves controlling these lines should then be locked open or shut, depending upon their function and position in the line. Air valves should be vented to the atmosphere; surge tanks and reservoirs may be drained. In any case, a buildup of residual pressure in the lines should be prevented. If lines are not equipped with lockout valves, it is recommended that they be installed. In some cases, it may be desirable to install two or more such valves on a line.

8. Check for mechanisms that are under spring tension or compression. Either block, clamp, or chain them in position.

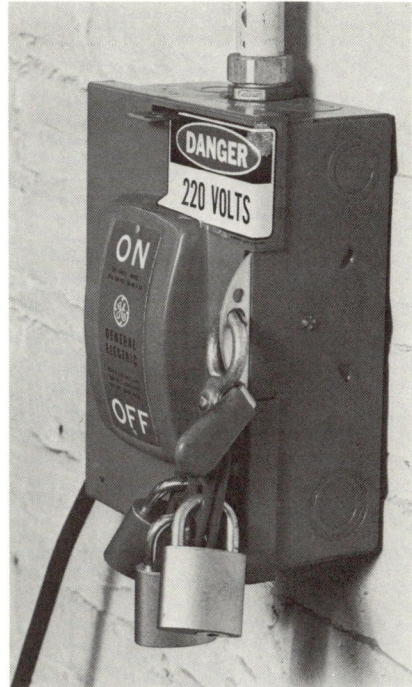

Figure 9-7.—Before working on a machine, the power must be turned off and locked out. Each worker should have his own individual padlock and key. Valves (*below*) should also be locked at Zero Energy State.

149

9. Check for suspended mechanisms or parts that normally cycle through a lower position and which could drop. Either lower them to their lowest position or block, clamp, or chain them in place.

10. Check for sharp or projecting parts or surfaces that could cut, tear, or gouge. Either remove or pad them, whichever is more convenient.

11. When the work is complete, each individual removes his padlock(s) and the last one removes the locking device and other safety devices. Never allow anyone to remove someone else's lock(s). Before removing safety devices and locks, be sure that other people will not be exposed.

If the key to a lock is lost, the owner should report it at once to his supervisor and get both a new lock and key. In some cases, equipment can be tagged out instead of locked out. Tags are not as effective as locks because tags are easily removed, overlooked, or ignored.

Normal maintenance and servicing should not be provided only on an emergency basis. Waiting until a machine breaks down can result in serious consequences. Good production cannot tolerate this approach for it is wasteful of time, money, and personnel.

Maintenance and service functions should be easy to schedule because they are already spelled out in the manufacturer's service manual. Additional functions should be added if changes have been made in this system.

Hand Tools

Because hand tools are so common and have been used by many people since they were children, they are often taken for granted. Not only are their importance and usefulness downgraded, but also their need to be used with ability and skill. In these days of automation and space-age wonders, hand tools are frequently thought of as being of less importance, something that can be abused and misused. This attitude causes many minor accidents along with a sizable number of serious ones.

In too many firms, little or no attention is paid to educating and training people in the use of hand tools. Too often it is assumed that employees possess these skills when they walk in the door.

Just as principles of safeguarding can be developed for machines and equipment, they can also be developed for hand tools. First they will be listed; then discussed one-by-one.

Know the correct tool for the job
Know how to use the tool
Use the correct tool for the job
Keep tools in good condition
Use needed personal protective equipment
Store tools properly

Know the correct tool for the job

Knowing just which is the best tool for doing a job can be a problem in view of the large number of tools that have been developed for special jobs. Most of these, however, fall into familiar

150

Figure 9-8.—Full-hand metal-mesh glove and apron reduces knife injuries.

categories—screw drivers, chisels, wrenches, pliers, knives, hammers, and the like. Unfortunately, even though the names are clear as to their intended use, they are often misused. For example, a screw driver is intended for assembly operations, not for use as a chisel, hammer, or pry bar. This abuse may be due to ignorance, but is more likely due to the unavailability of having the right tool when it is needed.

Know how to use the tool

Knowing how to use the tool properly is the product of training and experience. Some tools are simple; others complicated. Some simple tools, like a chisel, require experience to develop the right "feel" for doing the work properly.

Use the correct tool for the job

When planning a job, be sure to plan to use the correct tools. Maintenance people, for example, may check only the location of a job assignment; they may arrive only to find that they did not bring the proper tools. If it is a rush job or if it is a long way to go back, they may improvise.

Keep tools in good condition

If tools are used in a proper way, they usually remain in good condition. Cutting edges must be kept sharp. Moving parts must be kept lubricated and free from rust. If a tool cannot be restored to good condition, it should be discarded.

Use needed personal protective equipment

Using personal protective equipment can eliminate some real hazards. A hammer can send particles flying through the air—eye and face protection can thwart them. Cutting edges of some tools can be hazardous to fingers, hands, or arms—hand and body protection can be used. (See Figure 9-8). Wearing a safety hat is generally good practice when working about the plant.

Store tools properly

Many times, tools are taken to the job site and left there. If they are not stolen, they are usually damaged by water, dirt, or neglect. All tools should have a designated storage area; this can range from a sturdy box to a centralized tool crib.

Centralized tool control. A centralized tool control program will

151

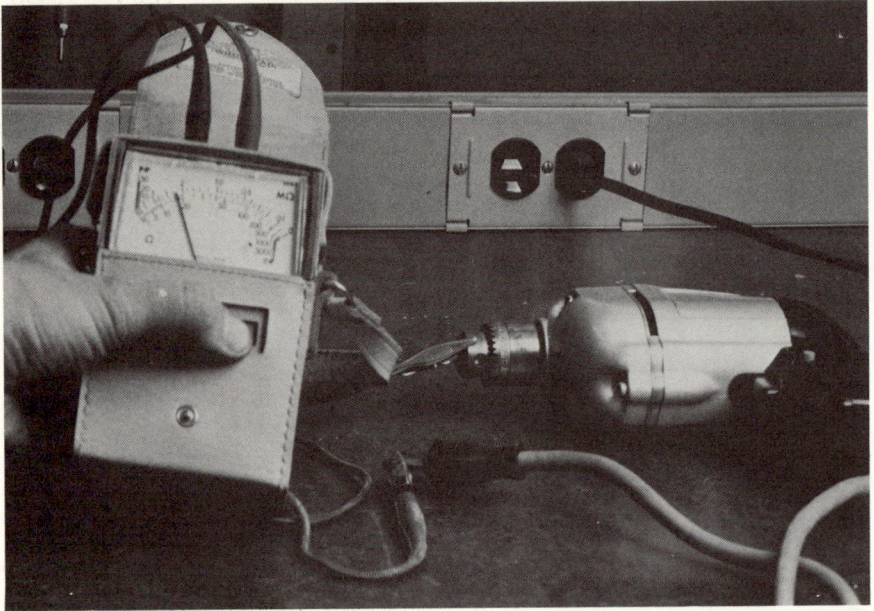

Figure 9-9.—Tools should be inspected and maintained by a trained person. Here insulation resistance of a portable tool is being tested.

help to assure effective inspection and maintenance of tools by a trained person. A competent tool room attendant can promote tool safety by issuing the right tool, encouraging employees to turn in defective or worn tools, and encouraging the safe use of tools. The required personal protective equipment can be issued when the tool is distributed.

The tool room attendant or tool inspector should be qualified by training and experience to pass judgment on the condition of tools for further use. Dull or damaged tools should not be returned to stock for reissuance. Enough tools of each kind should be on hand so that when a defective or worn tool is removed from service, it can be replaced immediately with a safe tool.

A procedure should be set up so that the tool supply room attendant can inspect and send tools in need of repair to the appropriate department or manufacturer for a thorough reconditioning. Issuance of numbered checks to employees permits the tool room attendant to know where each tool is and allows him to recall it for inspection at regular intervals.

With scattered work locations, it is not always practical to maintain a central tool supply room. In such cases, the job supervisor should inspect all tools frequently and remove from service those

152

found to be defective. A checklist can help facilitate inspections.

Portable Power Tools

In general, everything that was said about proper use and storage of hand tools applies to portable power tools. Power tools, however, have one major difference—an additional energy source. This source of power is usually electricity, but the tool may be powered by compressed air, gasoline, flexible shaft, or even be explosive actuated. Each will be discussed in turn.

Electric power tools

Under the wrong conditions, electricity can be a dangerous friend. Because electricity can cause serious injury and even kill, electric power tools should be checked daily for external signs of damage. (If used only occasionally, then they should be checked before each use.) Broken insulation, frayed cords, lack of ground fitting, and damage to the case are all good reasons not to use a tool. Because internal defects can also develop (usually through overloading or overheating the tool), all electric tools should be periodically tested with equipment that can detect internal faults.

Pneumatic tools

Air or pneumatic tools should be checked for constancy of rated rpm. Occasionally valves stick or are blocked so that rpm's soar above the rating of the tool and cause it to break apart.

Gasoline- and explosive-powered tools

Gasoline is flammable and should be dispensed with care and under controlled conditions. Explosive-actuated tools are nothing more than modified guns and should be handled as such. Cartridges should be handled as ammunition.

All portable power tools are, in effect, "walking machines." They have as much speed and power as many of their stationary counterparts. Many also need point-of-operation guarding. Controls should be safeguarded to prevent inadvertant actuation and they should be of the constant-pressure type which snap off when released.

Education and training of power tool operators is important, not only to develop safe and efficient skill, but also to develop respect and understanding of the equipment. Many people equate portable power tools with hand tools and minimize the additional hazard created by power. Another important reason for training is that people often use portable power tools on jobs remote to the regular work areas. Often the operator is alone and an accident can have serious consequences if no one is around to help.

Welding Equipment and Operations

Welding equipment is usually portable and is used for a wide variety of jobs about the plant. Welding operations present a number of hazards not normally associated with hand tools—the major hazard being heat. In addition to the fire hazard of an open flame or sparks or slag, there are hazards of toxic fumes and gases

153

Figure 9-10.—Self-standing safety shield provides close-quarter protection, keeps other workers from being exposed to arc rays.

and infrared and ultraviolet rays. Finally, the power source, whether electric or gas, can be a source of electrical shock or explosive force.

Welding hazards can be controlled easily by use of suitable personal protective equipment and proper work procedures. Welding goggles or helmets must be used. When helmets are used, safety glasses must also be worn whenever secondary finishing is done on the work piece. Respirators, gloves, aprons, and other protection should be worn as the nature of the specific job warrants. Permanent welding stations should be equipped with good local exhaust systems; fans can be used for small jobs or jobs of short duration. Respirators should be considered only when other methods are not available or are inadequate. Respirators should be used with a positive air supply any time welding is done in small, confined areas.

In addition to open flame and heat, welding produces sparks and slag. All clothing should be fire-resistant and worn so that it will shed sparks and slag (for example, a trouser leg should *not* be tucked into a boot top). In some cases, leggings and arm protectors should be used. Helpers and other nearby workers also need protection. Placing welding curtains or shields around the workplace may be necessary to confine the sparks and eliminate the haz-

154

ard of ultraviolet radiation. (See Figure 9-10.)

Electric cords and gas cylinders, regulators, and hoses need special attention and care. Hoses and cords should never be allowed to cross aisleways so that they become tripping hazards or so that they can be damaged by traffic; they should be regularly checked for damage and promptly replaced if damaged. Cylinders should be transported on a welding cart and should be chained securely at all times. Regulators should be kept clean and free of all foreign matter, particularly oil, and be used on the correct cylinder. Nonsparking tools should be used when making changes.

Fire precautions

As stated at the beginning of this section, heat is the major hazard associated with welding. Each year, careless welding practices cause fires resulting in major damage. The number of people hurt in these fires is not great because many fires occur after business hours.

Fires due to welding usually occur because of the violation of one or both these principles:

• If the object to be welded or cut cannot be readily moved to a safe area, all flammable materials in the welding area must be moved out.

• If the object to be welded or cut cannot be readily moved and if all flammable materials cannot be removed, protective covers must be spread over these materials and a fire watch guard shall stay in the welding area for the duration of the welding operation and for a suitable interval after the completion of the work.

Neglecting to watch the welding area for a suitable interval after the work has been completed has allowed many fires to break out. Sparks and hot slag can land on combustible materials and smolder for a long time before they blaze up. Therefore, a fire watch should be maintained for at least one-half to three-quarters of an hour after completion of the work.

If welding must be done over wood floors, they should be swept clean, wet down, and then covered with asbestos blankets, metal, or other noncombustible covering. Pieces of hot metal and sparks must be kept from falling through floor openings onto combustible materials.

Sheet metal, flame-resistant canvas, or asbestos curtains should be used around welding operations to keep sparks from reaching combustible materials. Welding or cutting should not be permitted in or near rooms containing flammable liquids, vapor, or dust. Neither should it be done in or near closed tanks which contain (or have contained) flammable liquids, until the tanks have been thoroughly drained, purged, and tested free from explosive gases or vapors.

No welding or cutting should be done on a surface until combustible coverings or deposits have been removed. It is important that flammable dusts or vapors not be created during the welding operation. Fire extin-

155

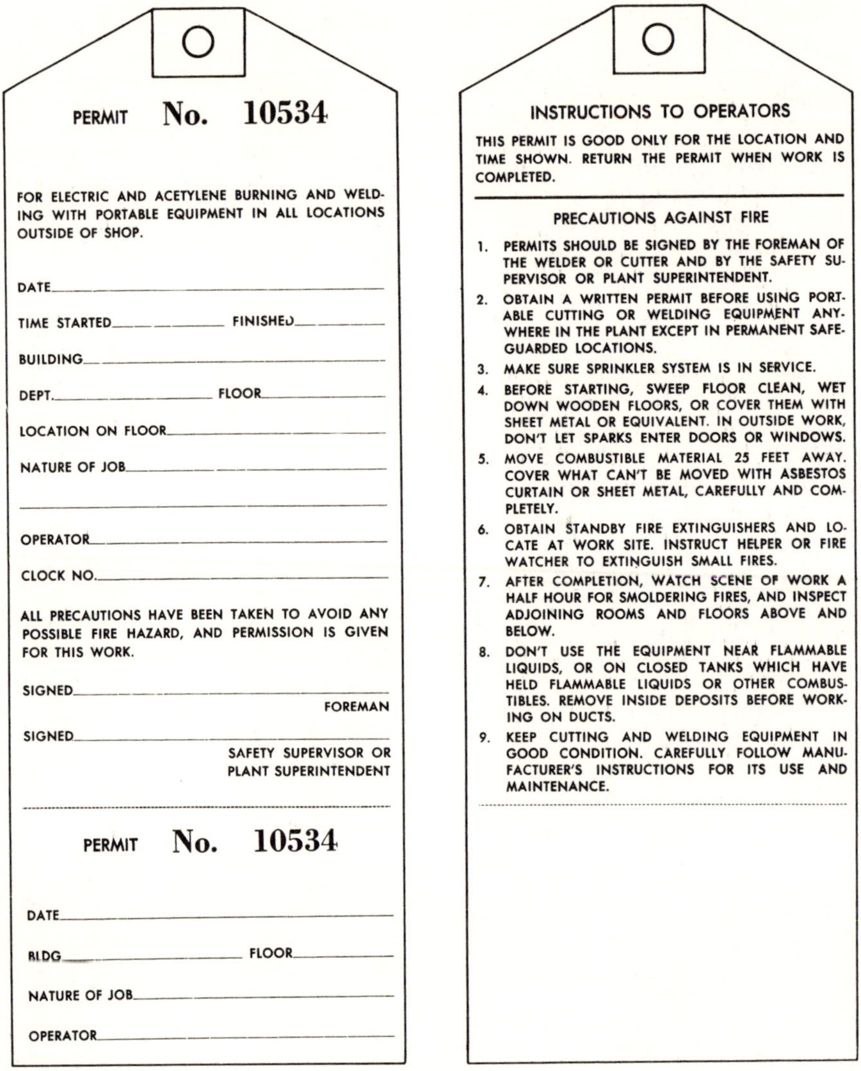

PERMIT No. 10534

FOR ELECTRIC AND ACETYLENE BURNING AND WELD-
ING WITH PORTABLE EQUIPMENT IN ALL LOCATIONS
OUTSIDE OF SHOP.

DATE_____

TIME STARTED_____ FINISHED_____

BUILDING_____

DEPT._____ FLOOR_____

LOCATION ON FLOOR_____

NATURE OF JOB_____

_____

OPERATOR_____

CLOCK NO._____

ALL PRECAUTIONS HAVE BEEN TAKEN TO AVOID ANY
POSSIBLE FIRE HAZARD, AND PERMISSION IS GIVEN
FOR THIS WORK.

SIGNED_____
                                        FOREMAN

SIGNED_____
                        SAFETY SUPERVISOR OR
                        PLANT SUPERINTENDENT

PERMIT No. 10534

DATE_____

BLDG_____ FLOOR_____

NATURE OF JOB_____

OPERATOR_____

INSTRUCTIONS TO OPERATORS

THIS PERMIT IS GOOD ONLY FOR THE LOCATION AND
TIME SHOWN. RETURN THE PERMIT WHEN WORK IS
COMPLETED.

PRECAUTIONS AGAINST FIRE

1. PERMITS SHOULD BE SIGNED BY THE FOREMAN OF
   THE WELDER OR CUTTER AND BY THE SAFETY SU-
   PERVISOR OR PLANT SUPERINTENDENT.
2. OBTAIN A WRITTEN PERMIT BEFORE USING PORT-
   ABLE CUTTING OR WELDING EQUIPMENT ANY-
   WHERE IN THE PLANT EXCEPT IN PERMANENT SAFE-
   GUARDED LOCATIONS.
3. MAKE SURE SPRINKLER SYSTEM IS IN SERVICE.
4. BEFORE STARTING, SWEEP FLOOR CLEAN, WET
   DOWN WOODEN FLOORS, OR COVER THEM WITH
   SHEET METAL OR EQUIVALENT. IN OUTSIDE WORK,
   DON'T LET SPARKS ENTER DOORS OR WINDOWS.
5. MOVE COMBUSTIBLE MATERIAL 25 FEET AWAY.
   COVER WHAT CAN'T BE MOVED WITH ASBESTOS
   CURTAIN OR SHEET METAL, CAREFULLY AND COM-
   PLETELY.
6. OBTAIN STANDBY FIRE EXTINGUISHERS AND LO-
   CATE AT WORK SITE. INSTRUCT HELPER OR FIRE
   WATCHER TO EXTINGUISH SMALL FIRES.
7. AFTER COMPLETION, WATCH SCENE OF WORK A
   HALF HOUR FOR SMOLDERING FIRES, AND INSPECT
   ADJOINING ROOMS AND FLOORS ABOVE AND
   BELOW.
8. DON'T USE THE EQUIPMENT NEAR FLAMMABLE
   LIQUIDS, OR ON CLOSED TANKS WHICH HAVE
   HELD FLAMMABLE LIQUIDS OR OTHER COMBUS-
   TIBLES. REMOVE INSIDE DEPOSITS BEFORE WORK-
   ING ON DUCTS.
9. KEEP CUTTING AND WELDING EQUIPMENT IN
   GOOD CONDITION. CAREFULLY FOLLOW MANU-
   FACTURER'S INSTRUCTIONS FOR ITS USE AND
   MAINTENANCE.

Figure 9-11.—Front and back of a typical hot work permit. Tag should be tied or mounted at the work site.

guishing equipment should be provided at each welding or cutting operation as standby equipment. A water pump tank unit is recommended for Class A fires, and a dry chemical or carbon dioxide for Class B and C fires. A watcher should be stationed to prevent stray sparks or slag from starting fires or to immediately

156

extinguish fires that do start while they are small. The area should be under fire surveillance for at *least one-half hour after* welding or cutting has been completed since many fires are not detected as soon as they start. (See explanations of terms in Chapter 12, Fire Protection.)

OSHA regulations 29 CFR 1910.252(d) cover these requirements in depth.

Hot work permits

Many firms use a "hot work permit" system to assure safety in welding operations. Here is how the system works. An area supervisor is notified that welding is to be done in his work area. The supervisor surveys the area to see that the object to be welded or cut can be moved to a safe area; if not, he makes sure flammables are removed from the area insofar as possible, and that protective covers are spread over combustibles that cannot be moved. He certifies the area to be safe "as is" or safe with a fire watch.

If welding and cutting must be carried on in locations other than where specifically designated, the operations must not be performed until (*a*) the areas have been surveyed for fire safety by persons who know the hazards, (*b*) the necessary precautions to prevent fires have been taken, and (*c*) a "hot work" permit has been issued. This permit must not be "stretched" to cover an area or an item or a time not originally specified—no matter how small the job may seem nor how little time

may be required to do it.

In general, welding is not permitted in or near rooms containing flammable vapors, liquids, or dusts or inside tanks and other containers that have held flammable liquids, until all fire and explosion hazards have been eliminated. If welding is necessary in such locations, the area should be ventilated to prevent an accumulation of explosive vapors. Spray booths and ducts that may contain combustible deposits should also first be freed of these materials before welding is permitted in them.

Most of these foregoing practices are required by OSHA and can be found in Subpart Q of Safety and Health Standards for General Industry (Title 29, *Code of Federal Regulations,* Part 1910) along with some additional prohibitions. Be sure to consult the latest regulations.

Summary

This chapter has broadly sketched the basics of safe machine and equipment operation, and safe handling of tools and welding equipment. Specialized help should be sought where needed. Check manufacturers of your machines and equipment, your insurance carrier, and safety consultants or specialists. Of course, the National Safety Council can assist with much additional material and information to make your operations safer and more healthful to your employees.

Materials Handling and Storage

Manual handling of materials accounts for about 23 percent of all occupational injuries—these injuries are from every part of an operation, not just the stock room or warehouse. As an average, industry moves about 50 tons of material for each ton of product produced.

Strains and sprains, fractures, and bruises are the common injuries. They are caused primarily by unsafe work practices—improper lifting, carrying too heavy a load, incorrect gripping, failing to observe proper foot or hand clearances, and failing to wear proper protective equipment. The largest number of injuries occur to fingers and hands.

Materials handling is a function that almost every employee performs—either as his sole duty or as part of his regular work.

Manual handling of materials increases the possibility of injuries and adds to the cost of a product. To reduce the number of materials handling injuries and to increase efficiency, materials handling should be minimized by combining or eliminating operations, or reorganizing them. What materials handling must still be done should be done mechanically, insofar as possible.

To gain an insight into the materials handling injury problem, these questions should be asked about operating practices.

• Can the job be engineered so that manual handling will not be necessary? Or if necessary, at least will be reduced?

• What types of injuries do employees receive?

• Can employees be given handling aids, such as tote boxes, trucks, or hooks, that will make their jobs safer?

• Will protective clothing or other personal equipment help prevent injuries?

• Can the employees be given special training to help prevent handling injuries?

Lifting and Carrying

Before employees are assigned to jobs requiring heavy and/or frequent lifting, make sure they are physically suited for the job. A person's lifting ability is not necessarily indicated by his height or weight. In some cases, a small man can lift heavy objects safely, whereas a big, husky 250-pounder may not be able to do so without injuring himself. Some states have laws that apply to how much a woman may lift in continuous or repeated operations. It should be noted that OSHA has no manual lifting load limits.

The following points are the basic rules for lifting or moving objects.

1. Check the area around the object and the route to be covered.
2. Inspect the object for slivers, jagged edges, burrs, rough or slippery surfaces. Remove protruding nails, corner clips, baling wire, and related hazards.
3. Get a firm grip on the object.
4. Keep fingers away from pinch points, especially when setting down materials.
5. When handling lumber, pipe, or other long objects, keep hands away from the ends to prevent them from being pinched.
6. Wipe off greasy, wet, slippery, or dirty objects before trying to handle them.
7. Keep hands free of oil and grease.

Lifting and setting down are the first and last movements performed in handling materials. When done by hand, it is during these movements that most strains occur. It is important to train employees in proper lifting techniques if lifting injuries are to be reduced. These include:

1. Consider the size, weight, and shape of the objects to be carried. Do not lift more than can

be handled comfortably. If necessary, get help.
2. Never carry a load that you cannot see over or around. Make sure the path of travel is clear.
3. Follow the six-step lifting procedure described in Figure 10-1. (Setting down an object requires just the reverse procedure.)

The proper way to lift

Lifting is so much a part of many everyday jobs that most of us do not think about it. But it is often done wrong, with unfortunate results.

In some cases, gloves, hand leathers, or other hand protectors have to be used to prevent hand injuries. However, their use must be closely supervised if they are to be worn around moving machinery.

In other cases, a handle or holder can be attached to the object itself, such as, a handle for moving an auto battery, tongs for feeding material to metal-forming machinery, or a wicker basket for carrying control laboratory samples. Drum lifters are also available. (See Figure 10-2.)

Feet and legs sustain a share of materials handling injuries, the greatest share being to the toes. One of the best ways to avoid injuries is to have people wear foot protection—safety shoes, instep protectors, and ankle guards; see Chapter 11.

The eyes, head, and trunk of the body can also be injured. When opening a wire-bound or metal-

(Text continues on page 162.)

159

Proper Way To Lift

Lifting is so much a part of everyday jobs that most of us don't think about it. But it is often done wrong, with bad results: pulled muscles, disk lesions, or painful hernia.

Here are six steps to safe lifting.

1. Keep feet parted—one alongside, one behind the object.
2. Keep back straight, nearly vertical.
3. Tuck your chin in.
4. Grip the object with the whole hand.
5. Tuck elbows and arms in.
6. Keep body weight directly over feet.

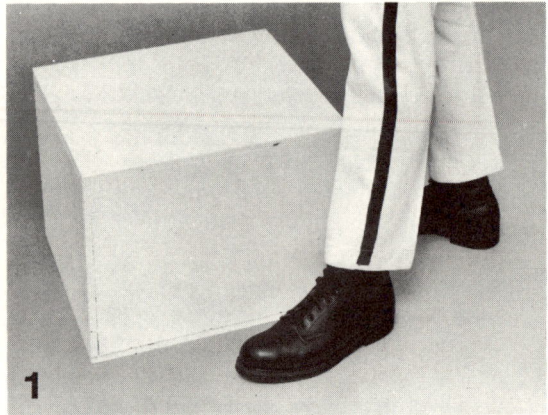

FEET should be parted, with one foot alongside the object being lifted and one behind. Feet comfortably spread give greater stability; the rear foot is in position for the upward thrust of the lift.

BACK. Use the sit-down position and keep the back straight —but remember that "straight" does not mean "vertical." A straight back keeps the spine, back muscles, and organs of the body in correct alignment. It minimizes the compression of the guts that can cause hernia.

Figure 10-1.

ARMS AND ELBOWS.
The load should be drawn close, and the arms and elbows should be tucked into the side of the body. When the arms are held away from the body, they lose much of their strength and power. Keeping the arms tucked in also helps keep body weight centered.

PALM. The palmer grip is one of the most important elements of correct lifting. The fingers and the hand are extended around the object you're going to lift. Use the full palm; fingers alone have very little power. Glove has been removed to show finger positions better.

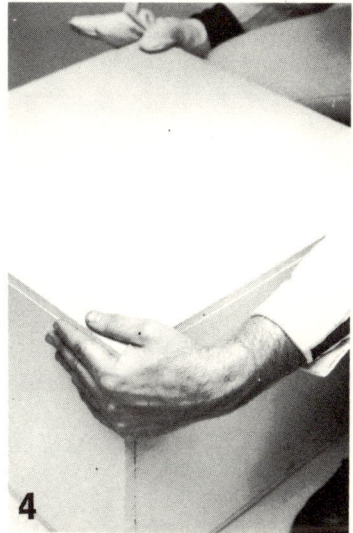

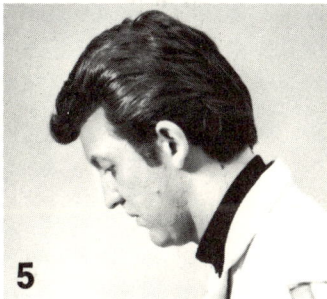

CHIN. Tuck in the chin so your neck and head continue the straight back line and keep your spine straight and firm.

BODY WEIGHT. Position body so its weight is centered over the feet. This provides a more powerful line of thrust and ensures better balance. Start the lift with a thrust of the rear foot.

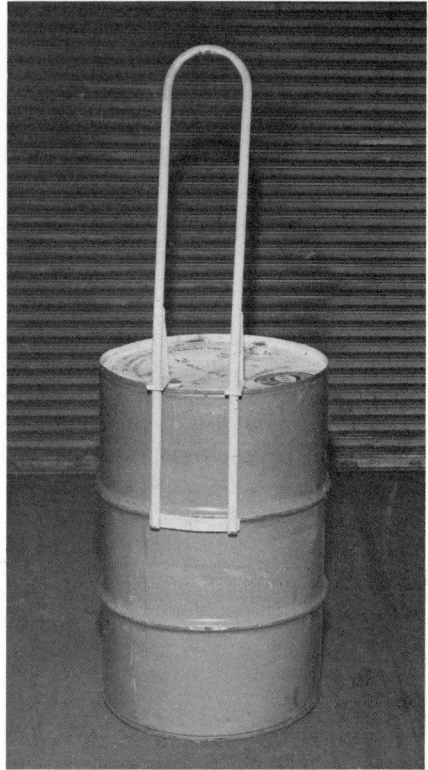

Courtesy Pickands Mather & Co., Hoyt Lakes, Minn.

Figure 10-2.—Two-handled lifter minimizes strain and provides safe control when lifting or lowering a drum.

bound bale or box, a person should wear eye protection, as well as stout gloves, and take special care to prevent the ends of the bindings from flying loose and striking the face or body. The same precaution applies to coils of wire, strapping, or cable.

If material is dusty or is toxic, the person handling it should wear a respirator or other suitable personal protective equipment.

Hand trucks

When not in use, hand trucks should be stored in a designated area, not left in aisles or other places where they would be a tripping hazard or traffic obstruction. Trucks with drawbar handles should be parked with handles up and out of the way. Two-wheeled trucks should be stored on the chisel with handles leaning against a wall or the next truck. Wheels of trucks not being used should be blocked.

Powered Materials Handling

Powered hand trucks

The use of the powered hand truck, controlled by a walking op-

162

erator, is increasing. The principal hazards of operating a truck in this manner are (*a*) the operator can be caught between the truck and another object, and (*b*) collisions with other objects.

The truck should be equipped with a dead-man control, wheel guards, and an ignition key that can be taken out when the operator leaves the truck.

No one should operate a truck unless he has had training and is authorized. Training should include the operating instructions given in the truck manufacturer's manual. General instructions include:

1. Do not operate the truck with wet or greasy hands.
2. Lead the truck from right or left of the handle. Face direction of travel. Keep one hand on the handle.
3. When entering an elevator, back the truck in to keep from getting caught between the handles and the elevator walls. Operate the truck in reverse whenever it must be moved close to a wall or other obstruction.
4. Always give pedestrians the right of way.
5. Stop at blind corners, doorways, and aisle intersections to prevent collisions.
6. Never operate the truck faster than normal walking pace.
7. Handle flammable or corrosive liquids only when they are in approved containers.
8. Never ride the truck, unless it is specifically designed for the driver to ride.

9. Never permit others to ride on the truck.
10. Do not indulge in horseplay.

These general rules for operators of powered hand trucks should also be followed by operators of powered industrial trucks.

Powered industrial trucks

Only physically qualified operators, who have received training in safe operation and who are duly authorized, should be permitted to drive any powered industrial truck.

According to OSHA requirements, training programs should include safe operating practices, as well as actual supervised experience driving over a training course. Emphasis should be on safety awareness. Trained and authorized drivers should have badges or other visible identification of authorization to drive, and these should be worn at all times.

Powered industrial trucks have either a battery-powered motor or an internal combustion engine. Trucks should be maintained according to their manufacturer's recommendations. Vehicle modifications which could affect the capacity and safe operation should not be performed by the user without the manufacturer's prior written approval. Eleven different designations of trucks or tractors are authorized for use in various locations. (For details, see OSHA regulations 29 CFR 1910.178, which is promulgated from ANSI B56.1-1969 and NFPA No. 505-1969.)

Battery-charging installations must be located in areas desig-

163

nated for that purpose. They must have facilities for flushing and neutralizing spilled electrolyte for fire protection, for protecting the charging apparatus from damage by trucks, and for adequate ventilation for dispersal of gases or vapors from charging batteries. Racks used for supporting batteries must be made of nonconductive material or be coated or covered to achieve this objective.

A carboy tilter or siphon must be used for handling electrolyte. Acid must always be poured into water; water must not be poured into acid (it overheats and splatters). Electrolyte can be purchased already mixed.

During charging operations, vent caps must be kept in place to avoid electrolyte spray. Make sure that vent caps are functioning. The battery compartment cover must be open to dissipate heat.

Precautions must be taken to prevent open flames, sparks, or electric arcs in battery-charging areas. Tools and other metallic objects must be kept away from the tops of uncovered batteries because the batteries may be short circuited.

Employees charging and changing batteries must be authorized to do the work. They must be trained in the proper handling techniques and required to wear protective clothing, including face shields, long sleeves, rubber boots, aprons, and gloves.

Signs prohibiting smoking in the charging area should be prominently displayed.

Refueling. All internal combustion engines must be turned off before refueling. Refueling should be done in the open or in specifically designated areas where ventilation is adequate to carry away fuel vapors.

General rules. No power truck should be used for any purpose other than that for which it has been designed. A forklift, for example, should not be used as an elevator for employees (for instance, in servicing light fixtures or when stacking materials) unless a safety pallet, with standard railing and toeboards, is fastened securely to the forks. Even then, no one should ride in the safety pallet when the lift truck is moving.

When internal combustion engine trucks are operated in enclosures, the concentration of carbon monoxide should not exceed the limits specified by local or state laws, and in no case should the time-weighted average concentration ever exceed 50 ppm (parts per million) for an 8 hour exposure. The atmosphere must contain a minimum of 19 percent oxygen, by volume. (Air usually contains 20.8 percent oxygen, by volume.) An industrial hygienist can check any areas that may be suspect.

Powered industrial trucks should have horns or other warning devices with identifiable sounds loud enough to be heard above plant noise, or lights which will provide comparable notice of the truck's presence. Trucks should stop at blind corners and at doorways before passing through them.

Some power trucks are designated for use in locations where

flammable gases or vapors are or may be present in the air in quantities sufficient to produce explosive or ignitable mixtures, or which are hazardous because of the presence of combustible dust or easily ignitable fibers or flyings. Only trucks approved and marked for such areas are to be permitted in them. (See OSHA regulations 29 CFR 1910.178 or NFPA Standard No. 505, *Powered Industrial Trucks,* for classification details.) At least one portable fire extinguisher approved for use on Class B and C fires (see Chapter 12) should be readily accessible.

Dock plates. Dock plates (bridge plates) should be fastened down in order to prevent them from "walking" or sliding when they are being used. Either the plate can be nailed down or holes can be drilled and pins or bolts dropped through them.

The wheels of trucks or trailers should be blocked or chocked to prevent movement when power trucks are at the dock for loading or unloading.

When not in use, dock plates should be stored in a safe area to avoid tripping hazards. They should be stored flat so they can't fall over.

Inspection. Power trucks should be inspected on a regular schedule. The operator should make a daily inspection of controls, moving parts and wheels. Defective brakes, controls, tires, lights, battery, lift system, steering mechanism and signaling equipment should be repaired before returning trucks to service.

Conveyors

The belt conveyor is the most common. All conveyors, however, should be designed and constructed to conform with applicable codes and regulations.

Gears, sprockets, sheaves, and other moving parts must be either protected by standard guards or positioned in such a way to eliminate personal injuries.

The entire conveyor mechanism should be inspected periodically and any part showing signs of excessive wear should be replaced immediately. Particular attention should be paid to brakes, backstops, antirunaway devices, overload releases, and other safety devices to make sure that all are operative and in good repair. The starter button or switch for a conveyor should be located so that the operator can see as much of the conveyor as possible. If the conveyor passes through a floor or wall, then each area should be equipped with starting and stopping devices and simultaneous operation of all starting buttons or switches should be required to start the conveyor.

Electrical or mechanical interlocking devices (or both types) should be provided which will automatically stop a conveyor when the unit has been stopped or is blocked so it cannot receive additional loads, or if the system is overloaded. Interlocks are also necessary where parts of conveyor may be opened for adjustments. Interlocks must stop the operation when opened. (See portion of Chapter 9 on safeguarding of machines.)

Where conveyors pass through

165

building floors, the openings should be guarded by standard handrails and toeboards. As a fire precaution, each opening should be protected against the passage of flame or superheated gases from one floor to the next by doors that close automatically or by fog-type automatic sprinklers, so placed as to provide a curtain of waterfog across the opening. Where a conveyor passes through a fire wall, similar protection should be provided.

Screw conveyor troughs should be completely enclosed with sectional covers which can be removed easily for inspection and maintenance. The covered sections should be interlocked so that the screw stops when a cover is lifted. Sometimes a heavy wire mesh screen is placed beneath the solid top cover to permit inspection and still guard the screw when the cover is not in place. (See Chapter 9 on safeguarding of machines.)

Where a conveyor passes over working areas, aisles, or thoroughfares, suitable safeguards shall be provided to protect those areas from the hazards of dislodged materials. Where conveyors are installed overhead and dip down at work stations, suitable guards or handrails shall be provided to prevent accidental injury to personnel.

Cranes and slings

Ropes, chains, and slings. All materials handling gear and equipment provided by the employer shall be inspected by the employer or his authorized representative before each use and when necessary, at intervals during its use, to make sure that it is safe. Any equipment which is found upon such inspection to be unsafe, must not be used until it is made safe.

Fiber rope. Precautions should be taken to keep rope in good condition. Kinking, for example, strains the rope and may overstress the fibers. Rope must be thoroughly dried out if it becomes wet, otherwise it will deteriorate quickly. Rope should not be stored or used in an atmosphere containing acid or acid fumes. Sharp bends should be avoided whenever possible because they cause extreme stress in the fibers. Rope should not be dragged because this causes abrasion.

Rope should be inspected at least every 30 days, more often if it is used to support scaffolding on which men work. The following procedure will assist in the inspection of rope:

• Check for broken fibers and abrasions.

• Inspect fibers by untwisting the rope in several places. If the inner yarns are bright, clear, and unspotted, the strength of the rope has been preserved to a large degree.

• Unwind a piece 8 inches long and $1/4$ inch in diameter from the rope. Try to break it manually. If it breaks with little effort, the rope is unsafe.

• Inspect rope that is used around acid or caustic daily. If black or rusty brown spots are noted, test the fibers (as just described) and discard all rope that

166

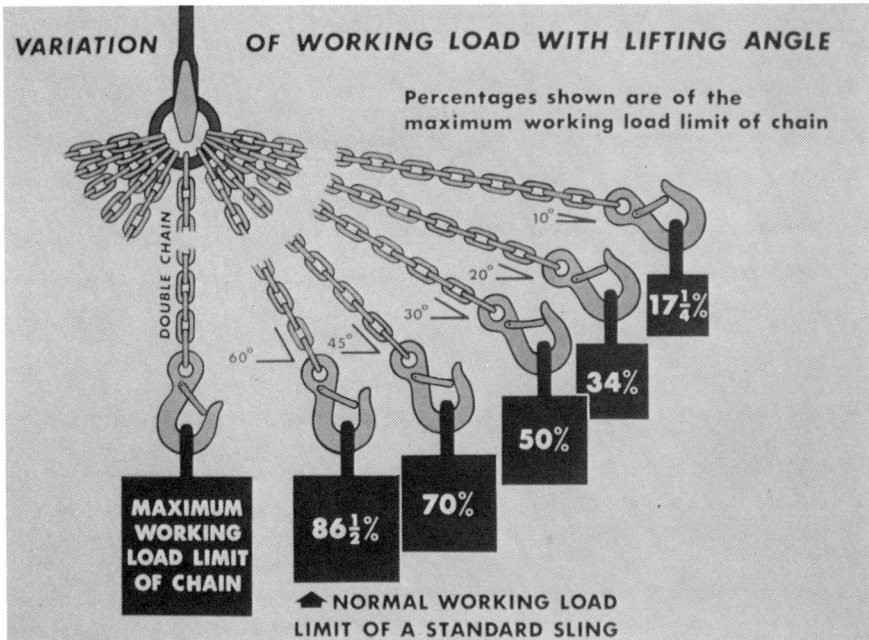

VARIATION OF WORKING LOAD WITH LIFTING ANGLE

Percentages shown are of the maximum working load limit of chain

DOUBLE CHAIN

10°
20°
30°
45°
60°

17¼%

34%

50%

70%

MAXIMUM WORKING LOAD LIMIT OF CHAIN

86½%

NORMAL WORKING LOAD LIMIT OF A STANDARD SLING

Figure 10-3.—Diagram shows how the angle of sling affects the working load limit.

fails the tests.

• Check eyes of slings; splices or end-fastenings must be in good condition.

Safe working loads of manila rope and rope slings are determined by the size of the rope and angle of sling. (See Figure 10-3. For specific details, see tables of manufacturer. General tables are given in NSC's *Accident Prevention Manual.*) As a general rule, a safety factor of not less than five should be maintained.

Synthetic fiber ropes. Where synthetic fiber ropes are substituted for manila ropes of less than 3 inches circumference, the substitute shall be of equal size. In making such a substitution it

should be ascertained that the inherent characteristics of the synthetic fiber are suitable for the intended service of the rope.

Wire rope and wire rope slings. Tables provided by the manufacturer shall be used to determine the safe working loads of various sizes and classifications of improved plow steel wire rope and wire rope slings having various types of terminals. For safe load limits, see the discussion in NSC's *Accident Prevention Manual.* OSHA's Safety and Health Regulations "Industrial Slings," 29 CFR 1910.184, has many tables.

The safe working load recommended by the manufacturer for

167

STEP 1.

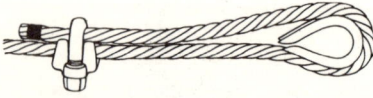

APPLY FIRST CLIP—one base width from dead end of wire rope—U-Bolt over dead end—live end rests in clip saddle. Tighten nuts evenly to recommended torque.

STEP 2.

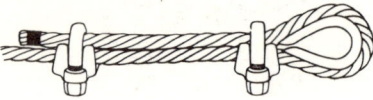

APPLY SECOND CLIP—nearest loop as possible—U-Bolt over dead end—turn on nuts firm but DO NOT TIGHTEN.

STEP 3.

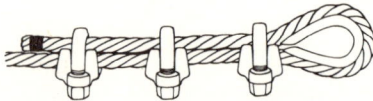

ALL OTHER CLIPS—Space equally between first two—this should be no more than one clip base apart—turn on nuts—take up rope slack—TIGHTEN ALL NUTS EVENLY ON ALL CLIPS to recommended torque.

Figure 10-4.—Method of clip installation.

specific, identifiable products shall be followed, provided that a safety factor of not less than five is maintained. The eyes of a sling must be formed or spliced in order to maintain the safe working load of the sling throughout. The type of end fastenings is also a factor in determining the safe working load of the sling.

Protruding ends of strands in splices on slings and bridles must be covered or blunted.

Where U-bolt wire rope fasteners are used to form eyes, the U-bolt shall be applied so that the U-section is in contact with the dead end of the rope (Figure 10-4). Clip fasteners shall be installed in the number recommended for the rope size:

$1/4$ to $5/8$ in.	3 clips
$3/4$ to 1 in.	4 clips
$1^1/8$ to $1^1/4$ in.	5 clips
$1^3/8$ to $1^1/2$ in.	6 clips

Clips shall be spaced a minimum distance of 6 times the rope diameter.

Wire rope slings shall be frequently inspected and lubricated. Blocks or heavy padding should be used at corners of the load to protect the sling from sharp bending and, when a multiple sling is used to lift a load, the sling shall be so arranged that the stress will be equalized between the ropes.

Wire rope shall be inspected before each use. Wire rope must not be used as load lifting gear if (*a*) in any length of eight diameters the total number of visible broken wires exceeds 10 percent of the total number of wires or if (*b*) the rope shows other signs of excessive wear, corrosion or defect.

Sheaves for rope should be as large as possible. The less flexible the rope, the larger the sheave or drum diameter must be; otherwise the rope will be bent too sharply.

Sheaves and drums should be aligned as much as possible to prevent excessive wear of rope. Reverse bending of rope, in which rope is bent first in one

168

direction and then in the other, should be avoided, as this wears out wire rope faster than anything else.

Fittings should be properly selected and attached if they are to be safe.

Wire rope should be spliced only by those who have been trained to do it properly. Splices should be tested to twice the load they are expected to carry.

Chains and chain slings. Alloy-steel chain, which is approximately twice as strong—size for size—as wrought iron chain, has become the standard material for chain slings. One advantage is that it is suitable for high-temperature operations.

An alloy-steel chain should never be annealed (when being serviced or repaired) because the process reduces the hardness of the steel, and this, in turn, reduces the strength of the chain. Wrought iron chain, on the other hand, should be annealed periodically by the manufacturer (or persons specially trained in the operation).

Impact loads, caused by faulty hitches, bumpy crane tracks, and slipping hookups, can materially add to the stress in the chain. Under a full working load, it will fail under impact before fully loaded wrought iron or heat-treated carbon steel chain.

Chain slings should preferably be purchased complete from the manufacturer, and whenever they need repair, they should be returned to their manufacturer. The manufacturer should also be

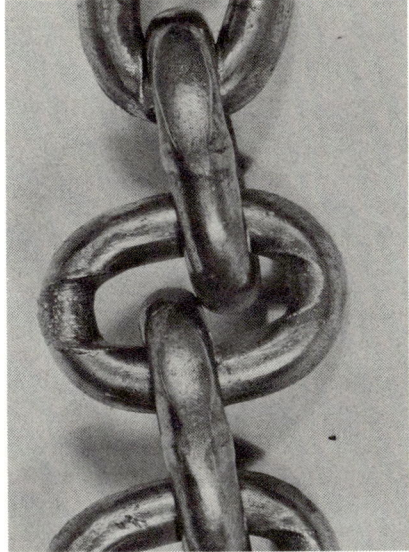

Figure 10-5.—Extreme wear at bearing surfaces. Links have been turned to detect wear.

queried about details of strength and specifications of the slings. The following recommendations must be followed:

• All sling chains, including end fastenings, shall be given a visual inspection before being used on the job. A thorough inspection of all chains in use shall be made every three months.

• Each chain shall bear an indication of the month in which it was thoroughly inspected.

• The thorough inspection shall include inspection for wear, defective welds, deformation and increase in length or stretch.

• Interlink wear, not accompanied by stretch in excess of five percent, shall be noted and the chain removed from service when

169

maximum allowable wear has occurred at any point of line. This varies from $3/64$ in. for $1/4$ in. chain, to $3/16$ in. for 1 in., to $11/32$ for $1^3/4$ in. (See Figure 10-5.)

• Chain slings shall be removed from service when, due to stretch, the increase in length of a measured section exceed 5 percent; when a link is bent, twisted or otherwise damaged; or when raised scarfs or defective welds appear.

• All repairs to chains shall be made under qualified supervision. Links or portions of the chain found to be defective shall be replaced by links having proper dimensions and made of material similar to that of the chain. Before repaired chains are returned to service, they shall be proof tested to the test load recommended by the manufacturer.

• Wrought iron chains in constant use shall be annealed or normalized at intervals recommended by the manufacturer. The chain manufacturer shall be consulted for recommended procedures for annealing or normalizing. Alloy chains shall not be annealed.

• A load shall not be lifted with a chain having a kink or knot in it. A chain shall not be shortened by bolting, wiring, or knotting.

• Standard tables shall be used to determine the maximum safe working loads of various sizes of wrought iron and alloy steel chains and chain slings; see 29 CFR 1926.251 and 29 CFR 1910.184.

Each chain should be tagged to indicate its load capacity, date of last inspection, date of purchase, and the type of material from which the chain is made. This information must not be stamped on the lifting links, however, as it may set up stress points that weaken the chain.

Chain attachments (rings, shackles, couplings, and end links) should be made of the same material to which they are fastened. Hooks should be made of forged or laminated steel and should be equipped with safety latches. Hooks should be replaced that have been overloaded, or loaded on the tips, and have a permanent set greater than 15 percent of the normal throat opening.

Hoists

Hoists—general. Scheduled detailed inspection of all hoists with special attention to load hooks, ropes, brakes, and limit switches is important. The safe load capacity of each hoist should be shown in conspicuous figures on the hoist body of the machine.

All hoists should be attached to their supports (fixed member or trolley) with shackels, or support hooks should be moused or have safety latches, see Figure 10-6. Latches are also recommended for load hooks.

Material hoists operating on rails, tracks, or trolleys should have positive stops or limiting devices either on the equipment, rails, tracks or trolleys to prevent overrunning safe limits and should be equipped with overspeed devices.

Extra protection against failure of the supporting hook, shackle,

170

Figure 10-6.—Hooks must be visually inspected daily. Hooks with cracks or having more than 15 percent in excess of normal throat opening or more than 10 degrees twist from the plane of the unbent hook must be replaced.

or block can be provided by a retaining cable or chain looped around the body of the hoist and the support (jib, beam, or carrying pin of the trolley).

A load should be picked up only when it is directly under the hoist. Otherwise, stresses for which the hoist was not designed may be imposed upon it. If the load is not properly centered, it may swing (upon being lifted), and injure someone. No one

should be under a raised load under any circumstance.

Unless used in combination with other safety devices approved by the hoist manufacturer, the simple floor-operated electric, air, or hand-operated hoists should not be used to lift, support, or otherwise transport human cargo. The standard commercial hoist or crane does not provide a secondary means of supporting the load should the wire rope or other suspension element fail.

Electric hoists. Control cords should have handles of distinctly different contours so that, even without looking, the operator will know which is the hoisting and which is the lowering handle.

Each control cord should be clearly marked "hoist" or "lower." Some companies attach an arrow to each control cord pointing in the direction in which the load will move when the rope is pulled. Also, it may be advisable to pass the control cords through a spreader to keep them from becoming tangled. The spreader can be a 1 by 3 inch board with equally spaced holes, resting upon the pull handles.

Control cords, usually made of fiber or light wire rope, should be inspected weekly for wear and other defects.

On pendant-controlled electric hoists, means for effecting automatic return to the "off" position should be provided on the control so that a constant pull on the control rope or push on the control button must be maintained to raise or lower the load.

171

Pushbutton control circuits should be limited to 110–120 volts. A limit stop should be installed on the hoist motion, and at least two turns of rope should remain on the drum when the load block is on the floor.

Air hoists. After a piston-type air hoist has been in operation for a time, the locknut that holds the piston on its rod may become loose so that the rod will pull out of the piston, thus letting the load drop. It is recommended that the locknut be secured to the piston rod by a castellated nut and cotter pin.

If an ordinary hook is used to hang the hoist from its support, the cylinder may come unhooked if the piston rod comes in contact with an obstruction when lowering. A clevis or other device should be used to prevent the hook from being detached from the hoist support.

Hand-operated chain hoists. Chain hoists should be permanently hooked onto a monorail trolley, or built into the trolley as an integral part. They are suitable for many operations on which a block and tackle fitted with fiber rope is used and are stronger, more dependable, and more durable than fiber rope tackle.

Screw-geared and differential hoists are self-locking and will automatically hold a load in position. Because the spur-geared type is free running, it tends to allow the load to lower itself. Therefore, an automatic load brake, similar to that on a crane, is provided to hold the load.

Storing Materials

General requirements

Both temporary and permanent storage must be neat and orderly. Materials piled haphazardly or strewn about increase the possibility of accidents to employees and of damage to materials.

Planned material storage reduces the handling necessary to bring materials into production, as well as facilitates the movement of the finished products from production to shipping.

When managers plan material storage, they should make sure that the materials will not obstruct fire alarm boxes, sprinkler system controls, fire extinguishers, first aid equipment, lights and electric switches and fuse boxes. All exits and aisles must be kept clear at all times.

There should be at least 18 inches clearance below sprinkler heads to reduce interference with water distribution. This clearance should be increased to 36 inches if the material being stored is very flammable.

A general rule of thumb is that aisles that carry one-way traffic should be at least 3 feet wider than the widest vehicle when loaded. If materials are to be handled from the aisles by a lift truck, the turning radius of the truck must also be considered. Employees should be told to keep materials out of the aisles and out of loading and unloading areas, which should be marked with guide lines.

Storage is facilitated and hazards are reduced by use of bins and racks. Material stored on

172

racks, pallets, or skids can be moved easily and quickly from one work station to another with less material damage and a reduced chance for employee injuries. When possible, material piled on skids or pallets should be cross-tied.

In an area where the same type of material is stored continuously, it is a good idea to paint a horizontal line on the wall to indicate the maximum height to which material may be stacked. This will help keep the floor load within the proper limits and the sprinkler heads will remain clear.

Containers and other material objects

Because packing containers vary considerably in size and shape and weight, height limitations for stacking these materials vary. The weight of the container itself must also be figured when keeping within the proper floor-loading limit.

Wire-bound containers should be placed so that sharp ends do not protrude into the passageway. When stacking corrugated paper cartons, remember that increased humidity may cause them to slump.

Sheets of heavy wrapping paper placed between layers of cartons will help prevent the pile from shifting. Better still, cross-tying will prevent shifting and sagging and permit higher stacking.

Bagged materials can be cross-tied with the mouths of the bags toward the inside of the pile, so contents will not spill into the aisle if the closure breaks. Piles over 5 feet high should be stepped back one row. Step back an additional row for each additional 3 feet of height. Sacks should be removed from the top of the pile, not half way up or from a corner. Sacks should be removed from an entire layer before the next layer is started. This keeps the pile from collapsing, possibly damaging material and injuring people.

Small diameter bar stock and pipe are usually stored in special racks so located that, when stock is removed, passers by are not endangered. The front of the racks should not face the main aisle, and material should not protrude into the aisles. When the storage area is set up, the floor load which this type of storage entails should be considered.

Larger diameter pipe and bar stock should be piled in layers separated by strips of wood or by iron bars. If wood is used, it should have blocks at both ends; if iron bars are used, the ends should be turned up. Again the floor load must be considered.

Lumber and pipe have a tendency to roll or slide. When workers make a pyramid pile, they should be told, therefore (a) to set the pieces down rather than to drop them, and (b) not to use their hands or feet to stop rolling or sliding material. These practices could cause serious injury.

Sheet metal usually has sharp edges and should be handled with hand leathers or leather gloves or gloves with metal inserts. Stacking should be in bun-

173

dles with the amount in each bundle varying according to the gage and size of the sheets. To facilitate handling and lessen the danger of shifting or sliding, the bundles should be separated by strips of wood, thick enough to permit the insertion of a forklift or other means for lifting.

Tin plate strip stock is heavy and razor sharp. Should a load, or partial load, fall, it could badly injure anyone in the way. Two measures can be taken to prevent spillage and injuries: (*a*) band the stock after shearing, and (*b*) use wooden or metal stakes around the stock tables and pallets that hold the loads. It is the responsibility of the supervisor and all who handle the bundles to make sure the load is banded properly and that the stakes are in place when the load is on a table.

Packing material, such as plastics, excelsior, or shredded paper, should be kept in a fire-resistant room provided with sprinklers and dust-proof electric equipment. Packing material received in bales should be kept baled until it is used, and then only enough for immediate use should be removed. If material is to be stored in the packing room, it should be placed in metal or metal lined bins with covers that are normally closed. In bins having counterweighted covers, the counterweight ropes should be equipped with fusible links to make certain that the covers close automatically in case of fire. As an added precaution against injury, in case the counterweight ropes break, the counterweights should

174

be boxed.

Barrels or drums stored on their sides should be pyramided with the bottom rows securely blocked to prevent rolling. If piled on end, the layers should be separated by planks. If pallets are used, they should be large enough to prevent overhang.

Hazardous liquids

Acid carboys are best handled with special equipment, such as carboy trucks. Boxed carboys should generally be stacked no higher than two tiers and never higher than three. Not more than two tiers should be used for carboys of strong oxidizing agents, such as concentrated nitric acid or concentrated hydrogen peroxide. The best method of storage would be in specially designed racks.

The safest way to draw off liquid from a carboy is to use a vacuum pump, or to use a syphon started by either a rubber bulb or an ejector. Another method is to use a carboy inclinator of the type that holds the carboy at the top, sides, and bottom, and returns the carboy to an upright position automatically when it is released.

Pouring by hand or starting pipettes or syphons by mouth suction to draw off the contents of a carboy should never be permitted.

When liquids are handled in portable containers, contents should be clearly labeled. Substances should be stored separately in a well-lit area so that people will not make mistakes when drawing liquids out.

All purchasing specifications should require that all chemicals be properly labeled. Fire and explosion hazard data, health hazard data, chemical reactivity, chemical and trade names, precautions for spills and leaks, physical property and special protection for storing and handling information should be provided by the manufacturer or distributor.

Drums containing hazardous liquids should be placed in a rack, not stacked. It is best to have a separate rack for each type of material being handled. Racks permit good housekeeping, easy access to the drums for inspection, and easy handling. Racks permit drums or barrels to be emptied through the bung.

Self-closing spigots should be used, particularly where individuals are allowed to draw off their own supplies. If there is a chance that spigots might be hit by material or equipment, drums should be stored on end and a pump used to withdraw the contents.

Drums containing flammable liquid, and the racks that hold them should be grounded, and the smaller container into which the drum is being emptied should be bonded to the chime of the drum by a flexible wire with a battery clip or C-clamp at each end. (See Figure 12-4, page 217.)

Storage areas for liquid chemicals should be well ventilated. Natural ventilation is better than mechanical, because a mechanical may fail. The floor should be made either of concrete or of some other resistant material treated to reduce absorption of the liquids. The floor should be

Courtesy Stauffer Chemical Co.

Figure 10-7.—Here's one way to test a safety shower. Note deluge quantity of water.

pitched toward one or more drains that are corrosion resistant and easily cleaned.

Where caustics or acids are stored, handled, or used, emergency flood showers of eyewash fountains should be available. (See Figure 10-7.) Workers should be provided with chemical goggles, rubber aprons, boots, and gloves, and other protective equipment necessary to handle the particular liquid. (See Chapter 11.)

If large quantities of liquid chemicals or solvents are used,

175

pipelines to outside storage tanks are safer because they reduce and localize any spillage.

When hazardous liquids are used in small quantities, only enough for one shift should be kept on the job. OSHA and local ordinances limit the amount of hazardous materials which can be stored or processed inside a building. The main supply should be stored in tanks located in an isolated place.

Pipelines should be color coded to identify their contents. Basic colors for identification of contents of pipes by class and methods of providing specific identification can be found in Chapter 3, under Use of Color. Should outside contractors work in the area, they should be adequately informed of piping systems and any hazards.

When a valve is to be worked on, it should be closed and the section of pipe in which it is located should be thoroughly drained before the flanges are loosened.

Tank cars should be protected on sidings by derails and by blue stop flags or blue lights before they are loaded or unloaded. Hand brakes should be set and wheels chocked. Before the car is opened, it should be bonded to the loading line.

The track and the loading or unloading rack should be grounded and all connections checked regularly.

Tank trucks should have the brakes set, wheel chocked, and grounding and bonding wires connected before unloading operations start.

Gas cylinders

Compressed gas cylinders should be stored on end on a smooth surface. All cylinders should be chained or otherwise fastened firmly against a wall, post or other solid object, or stored in racks. Different kinds of gases should be either separated by aisles or stored in separate sections of the building or storage yard. Empty cylinders should be stored apart from full cylinders.

To prevent cylinders from being run against or being knocked over, the storage areas should be removed from heavy traffic. Cylinders of flammable gases never should be stored near highly flammable substances. During winter, cylinders that become caked with ice and snow should be taken inside and allowed to thaw out gradually before opening them. Do not use a steam hose, or the heat may melt the fusible plug and release the cylinder contents. Do not use cylinders as rollers under any circumstances.

Handle all cylinders with care—a cylinder marked "empty" just might not be.

To transport cylinders, use a carrier that does not allow excessive movement, sudden and violent contacts, and upsets. When a two-wheeled truck with rounded back is used, chain the cylinder upright. Never use a magnet to lift a cylinder. For short-distance moving, a cylinder may be rolled on its bottom edge, but never dragged. Cylinders should never be dropped or permitted to strike one another. Protective caps must be kept on cylinders when cylin-

176

ders are not being used.

When in doubt as to how to handle a compressed gas cylinder or how to control a particular type of gas once it is released from the cylinder or other source, consult the supplier.

Combustible and dangerous solids

Bulk storage of grains and granular or powdered chemicals or other materials presents fire and explosion hazards. Many materials which are not considered hazardous in solid form often become quite combustible when finely divided. Some of these are carbon, fertilizers, food products and by-products, metal powders, resins, waxes and soaps, spices, drugs, and insecticides, wood, paper, tanning materials, chemical products, hard rubber, sulfur, starch, and tobacco. Though it is long, this list is by no means complete—it only shows the general categories of explosive dusts.

To avoid dust explosions, prevent the formation of an explosive mixture or eliminate all sources of ignition. Either the dust must be kept down, or enough air supplied to keep the mixture below the combustible limit.

Good housekeeping and good dust-collection equipment will go far toward preventing disaster. Formation of layers of dust on floors and on other structural parts of a building should be prevented. A dust explosion is a series of explosions, or a progression of explosions. The initial explosion is very small, but it shakes up dust that has collected on rafters and equipment so that a second explosion is created. This prog-ress usually continues until the result is the same as if a large explosion took place.

Wearing protective equipment is important when dust is toxic. Such equipment may range from a respirator to a complete set of protective clothing. A state of local governmental health and safety agency can recommend the proper type of protective clothing for a specific job. Also see the recommendations in Chapter 11, Personal Protective Equipment.

Bins in which solids are stored should have a sloping bottom to allow material to run out freely and to prevent arching. For some materials, especially if they are in-process, the bin should have a vibrator or agitator in the bottom to keep the material flowing.

When entering larger rectangular bins with power shovels or other power equipment, care should be taken not to undercut the material, which is extremely dangerous if arched material should suddenly give way. Where practical, tops of bins should be covered with a 2 inch mesh screen, or at least a 6 inch grating (or parallel bars on 6 inch centers) to keep people from falling into the bins.

Tests for oxygen content and for presence of toxic materials should be made before anyone enters a bin or tank for any purpose. If such tests indicate the need for protection, men should don air supplied masks, in addition to a safety belt and lifeline, before entering. See Chapters 11 and 13 of this handbook for details on personal protective equipment and work in hazardous and/or confined locations. The filling equipment should be made inoperative so that it cannot be started again except by the worker after he leaves the bin or by his immediate supervisor after he has checked the worker out of the bin. When a person is working in a bin, he should have a companion employee, equipped to act in case of emergency, stationed on top of the bin. Bins should be entered from the top only.

If flammable dusts, vapors, or other flammable air contaminants are present, workers should use only spark-resistant equipment and electrical equipment that is listed by Underwriters Laboratories Inc. for such hazardous locations.

OSHA also requires that the employer instruct employees who will be exposed to hazardous materials as to the nature of the hazards and the means of avoiding the hazards.

Shipping and Receiving

Floors, ramps, and aisles

Floors in warehouses, storerooms, and shipping rooms must be strong enough to allow no appreciable sag. Uneven floors may lead to the toppling of piles of stored materials. (Also see references to floors in Chapters 3 and 4.)

Wherever materials are stored or transported, the surface of floors, platforms, and ramps should be kept in good condition. Breaks should be repaired immediately, and the area around doorways and elevator entrances should particularly be watched.

Ramps should have nonskid surfaces. When ramps are used for hand trucking, a nonskid foot strip may be laid in the center, or in the center of each lane for two-way traffic. Ramps should have handrails.

Good housekeeping contributes to safety in handling of material, whether by hand or mechanically. A fall by a person carrying an object might result in a more serious injury than if his hands were free.

Mirrors, placed at blind intersections, help to prevent collisions. Warning signs and signals at such locations also serve as useful reminders, particularly to operators of power equipment. Doorways and entrances to tunnels and elevators may be similarly protected.

Mobile equipment used in storage areas should be equipped with backup warning devices.

Aisles should be wide enough to enable employees to move about freely, while handling material or removing it from bins, racks, or piles, and to allow safe passage of loaded equipment.

Lighting

General illumination of warehouses and shipping rooms should be at least 10 foot-candles for rough bulky materials, and 20 foot-candles for medium materials, and 50 foot-candles for fine materials requiring care. Special lighting should be provided for operations requiring greater illumination. All lighting fixtures and wiring should be in accordance with the requirements of the NFPA *National Electrical Code,* Standard No. 70.

178

Personal Protective Equipment

CHAPTER 11

Controlling the Hazard

Engineering and administrative controls

Engineering controls. When a situation that endangers the safety and health of the employees exists, the employer must immediately take steps to eliminate it. Extensive engineering revision of the processing or manufacturing methods may be involved or only a simple change in materials handling methods may be necessary. As described in Chapter 9, a machine safeguard that effectively confines flying particles eliminates a cause of accidents. This is a more basic treatment of the problem than the use of goggles designed to prevent injury, for confinement stops particles at their source.

Likewise, dangerous solvents, chemicals, and other vapor or fume hazard substances should be confined to a pipe or closed tank, or their vapors should be exhausted mechanically, or nonhazardous chemicals can be substituted, instead of dependence being placed on a respirator to protect an operator required to work in a hazardous environment.

Engineering controls—protection by mechanical means or process design—is generally more reliable than protection dependent upon human behavior.

Administrative controls. In the case of exposure to air contaminants and noise, the employer may use administrative controls, such as rotation of assignments and/or limiting the amount of time an employee will work at an operation so that his exposure to harmful air contaminants or physical agents will not exceed the permissible levels of exposure established for the materials or agent. More detailed discussions will be found in Chapter 13.

Personal protective equipment

When the hazard cannot be eliminated through engineering revisions, controls, guarding, or through administrative procedures, the use of personal protective equipment is mandatory.

General requirements. OSHA regulations require that protective equipment (including personal protective equipment for eyes, ears, face, head, and extremities), protective clothing, respiratory

179

devices, and protective shields and barriers shall be provided, used, and maintained in a sanitary and reliable condition wherever it is necessary by reason of hazards of processes or environment, chemical hazards, radiological hazards, or mechanical irritants encountered in a manner capable of causing injury or impairment in the function of any part of the body through absorption, inhalation, or physical contact.

Employee-owned equipment. Where employees provide their own protective equipment, the employer shall be responsible to assure its adequacy, including proper maintenance and sanitation of such equipment.

Design. All personal protective equipment shall be of safe design and construction for the purpose for which it is intended.

Basic program guidelines. To have an effective personal protective equipment program that will meet general requirements, the employer and/or the person responsible should:

1. Be familiar with required standards and the Occupational Safety and Health Act requirements
2. Be able to recognize hazards
3. Be familiar with the best safety equipment available to protect against these hazards
4. Know the procedures for supplying the equipment
5. Know how to maintain and clean equipment

6. Develop an effective method for training and persuading employees to use protective equipment when required

Equipment selection and use

Selection. Once the need for personal protective equipment has been established, the employer has the problem of selecting the proper type. Two criteria should be used: (*a*) the degree of protection which a particular piece of equipment affords under varying conditions, and (*b*) the ease with which it may be used.

Satisfactory performance specifications exist for certain types of personal protective equipment, notably respiratory devices, safety hats, devices to protect the eyes from impact and from harmful radiations, and rubber insulating gloves, but there are few approving laboratories to test equipment regularly according to these specifications. Unless the employer has testing facilities available to him, he must rely upon the equipment manufacturers' endorsement for devices he selects to fill his needs.

Equipment approvals. When selecting or buying any type of safety equipment, insist on the best equipment and deal only with reputable firms and specify required approvals. To assure the receipt of acceptable protective equipment, the purchasing contracts should stipulate the type of equipment, its intended use, and the specifications of the standard or regulation (ANSI, MESA, NIOSH) it must meet. Approved respiratory protective equipment

180

meet the requirements of NIOSH (National Institute of Occupational Safety and Health) and MESA (Mining Enforcement Safety Administration).

Some of the equipment standards require that the equipment bear markings, labels, or identification to indicate that the equipment is approved, listed, tested, etc., and safe for the stated use.

Proper use of equipment. The next major problem facing the employer is making certain that workers use personal protective equipment properly once it has been chosen. Several factors influence the solution of this problem. Among them are: (*a*) the extent to which people who must wear the equipment understand its necessity, (*b*) the ease and comfort with which it can be worn with a minimum of interference with normal work procedures, and (*c*) the available economic, social, and disciplinary sanctions which can be used to get people to use the equipment.

In job where workers are accustomed to wearing personal protective equipment as a condition of employment, this problem is minor. When people are issued personal protective equipment for the first time, the problem may be more difficult. Employees must be given a clear and reasonable explanation as to why the equipment must be worn. The practice of having supervisors try out new protective equipment prior to actual adoption, and getting their reactions, has been successfully used in many operations. Another successful technique is to have members of the safety committee or work force help select approved devices, use them for a trial period, and get their reaction.

Policies differ with respect to who pays for personal protective equipment. The cost of eye protection equipment that has prescription lenses, such as safety spectacles, is often shared by worker and employer. Most companies do not furnish safety toe shoes free. Some arrange for a shoe store or shoemobile service so employees can purchase the shoes through them. Often a company will reimburse the employee for part of the purchase price.

No matter what the policy may be concerning "who pays," it remains the employer's responsibility to enforce the use of required protective equipment.

Common Body Hazards and Protection

Head protection

General protection required. Helmets for the protection of occupational workers from impact and penetration from falling and flying objects and from limited electric shock and burn shall meet the requirements and specifications established in American National Standard *Safety Requirements for Industrial Head Protection,* Z89.1. (OSHA regulations adopt the ANSI standard.)

Safety hats. Safety hats (or helmets) are rigid headgear designed to protect the workers—not only from impact but from flying particles and electric shock or any combination of the three.

Helmets (see Figure 11-1) have

181

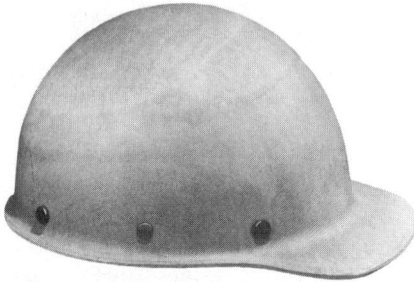

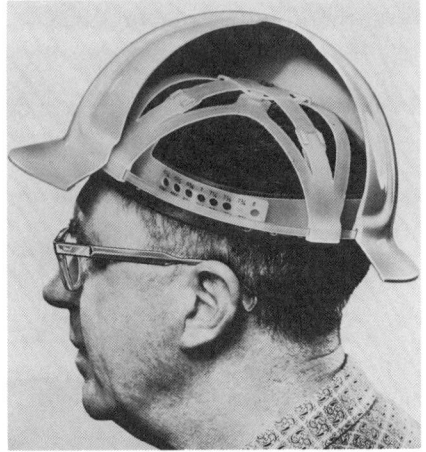

Figure 11-1.—*Left:* Protective helmet (cap style), molded of reinforced plastic. *Right:* Cutaway of full-brimmed helmet shows how crown straps are set to assure clearance for impact protection. Underside of hat should be no less than 1¼ inch above the head.

been classified into two types: (*a*) full brimmed, and (*b*) brimless with peak. The types have been further broken down into four classes: Class A, limited voltage resistance for general service; Class B, high voltage resistance; Class C, no voltage protection (metallic helmets); and Class D, limited protection for fire fighting service.

All helmets are identified on the inside of the shell with the manufacturer's name, American Standard designation Z89.1, and class (A, B, C, or D).

The headpieces designed for use around electrical hazards must meet performance voltage requirements of 20,000 volts AC, while the other classes have less strict electrical hazard requirement (2200 volts AC).

All helmets are designed to withstand a maximum average

force of up to 850 pounds.

A brim around the helmet as shown in Figure 11-1 provides the most complete protection for head, face, and back of neck. For use where a brim may be in the way, the cap type is frequently preferred. This type may also be equipped with lugs to support a welding mask for welders who work on jobs where there is a danger of heat injury.

Metal helmets. Metal helmets do not afford the high impact resistance of plastic ones, but because of their lighter weight, they are preferred by some workers. Metal helmets should *never* be used where electrical harzards or corrosive substances are present.

Suspensions, liners, and chinstraps. It is the suspension that gives a helmet its impact-distributing ability (Figure 11-1).

182

It is important that the suspension be adjusted to fit the wearer and keep the hat itself a minimum distance of 1¹/₄-inch above the wearer's head. Suspension bands should be nonirritating to the wearer. Liners are available so that protective helmets may be worn in comfort in cold weather. Various types of leather, fabric, and elastic chinstraps are available. Often hats are bumped or blown off or fall off. A chinstrap affords full protection; a nape strap (provided with most helmets) helps keep the headgear from falling off during normal use.

Bump caps. Another form of headgear is the bump cap, made of thin-shelled, lightweight plastic. Some industries have adopted them and they are new enough so that, at this time, there are no specifications covering them. But authoritites warn that, although they are fine for some applications, they are *not* a substitute for helmets. Use of such caps should be strictly limited.

Caps for long hair. It is important that those with long hair who work at machines protect their hair from contact with moving parts. Besides the danger of direct contact with the machine, the hazard of having hair lifted into moving belts or rolls by the heavy charges of static electricity which are developed should also be considered. Where such hazards exist, women and men with long hair should be required to wear protective hair covering (Figure 11-2).

Hair nets, bandannas, and tur-

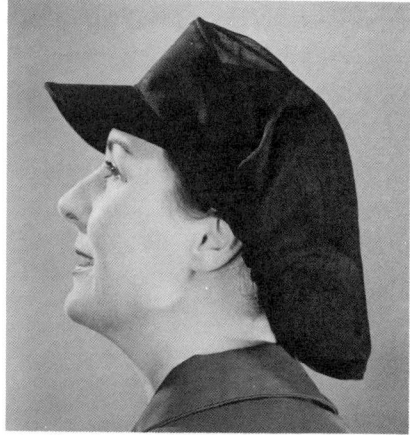

Figure 11-2.—Cap protects long hair in general factory work. The visor serves as a feeler guard to prevent injuries from contact with machines and objects. The netting construction on the top provides adequate comfort ventilation.

bans are frequently unsatisfactory for hair protection because they do not cover the hair completely. Protective caps should completely cover the hair. If the wearer is exposed to sparks and hot metals, as in spot welding, the cap should be made of flame-resistant material. Disposable flame-proof caps are provided in some chemical plants.

Although no standards have yet been accepted, caps should be made of a durable fabric to withstand regular laundering and disinfecting. Design should be simple—for ironing or pressing—and should be available in a variety of head sizes or adjustable to fit all wearers. A cap should have a visor long enough and rigid enough to provide warning before the head itself comes into contact with a moving object, such as the

183

spindle on a drill press. The cap should be as attractive as possible, cool, and lightweight. If dust protection is not required, the cap should be made of open-weave material for better ventilation.

Maintenance of equipment. Before each use, helmets should be inspected for cracks, signs of impact or rough treatment, and wear that might reduce the degree of safety originally provided. Once damaged, a protective helmet should be discarded. Alterations of any sort impair the performance of the headgear.

Before reissuing used helmets to other employees, they should be scrubbed and disinfected.

The condition of the suspension should be carefully inspected because of the important part it plays in absorbing the shock of a blow. Look for loose or torn cradle straps, broken sewing lines, loose rivets, defective lugs, and other defects. Sweatbands are easily replaced.

Many companies make use of colored safety hats to identify different working crews. Some colors are painted on and others have the color molded in. Before painting a hat, consult with its manufacturer so that a coating can be chosen which will not reduce the dielectric properties or attack and soften the shell material. Light-colored hats are cooler to wear in the sun or under infrared energy sources.

Eye protection
Hazards and requirements.

Hazards—exposures. Industrial operations expose the eyes to a variety of hazards—flying objects, splashes of corrosive liquids or molten metals, dusts, and harmful radiation are common examples. Eye injuries not only disable a person, but they often disfigure him. Per-injury cost is high to both employee and employer. Flying objects cause most eye injuries—metal or stone chips, nails, and abrasive grits. Corrosives and harmful substances, hot splashing metal, and light and heat rays can also damage the eye.

General requirements. Federal regulations stipulate that protective eye and face equipment shall be used where there is reasonable probability of injury that can be prevented by such equipment. In such cases, employers shall make conveniently available a type of protector suitable for the work to be performed, and employees shall use these protectors. No unprotected person shall knowingly be subjected to a hazardous environmental condition. Suitable eye protectors shall be provided where machines or operations present the hazard of flying objects, glare, liquids, injurious radiation, or a combination of these hazards.

Performance and design requirements. Eye and face protectors shall meet the following minimum requirements. They shall:

1. Provide adequate protection against the particular hazards for which they are designed

2. Be reasonably comfortable when worn under the designated conditions

3. Fit snugly and shall not unduly interfere with the move-

184

SELECTION OF EYE AND FACE PROTECTIVE EQUIPMENT

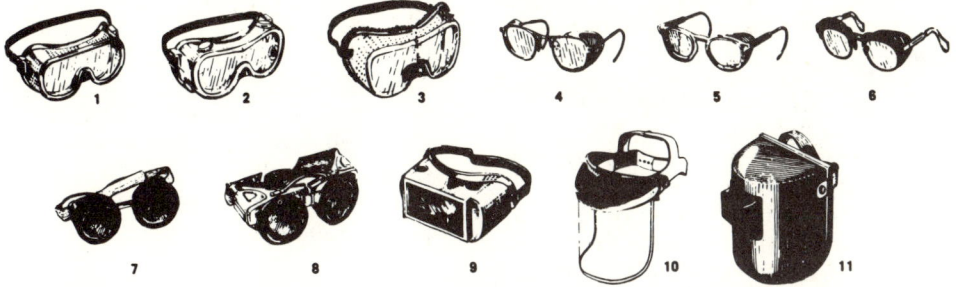

1. **GOGGLES**, Flexible Fitting, Regular Ventilation
2. **GOGGLES**, Flexible Fitting, Hooded Ventilation
3. **GOGGLES**, Cushioned Fitting, Rigid Body
*4. **SPECTACLES**, Metal Frame, with Sideshields
*5. **SPECTACLES**, Plastic Frame, with Sideshields
*6. **SPECTACLES**, Metal-Plastic Frame, with Sideshields

** 7. **WELDING GOGGLES**, Eyecup Type, Tinted Lenses (Illustrated)
7A. **CHIPPING GOGGLES**, Eyecup Type, Clear Safety Lenses (Not Illustrated)
** 8. **WELDING GOGGLES**, Coverspec Type Tinted Lenses (Illustrated)
8A. **CHIPPING GOGGLES**, Coverspec Type, Clear Safety Lenses (Not Illustrated)
** 9. **WELDING GOGGLES**, Coverspec Type, Tinted Plate Lens
10. **FACE SHIELD** (Available with Plastic or Mesh Window)
11. **WELDING HELMETS

*Non-sideshield spectacles are available for limited hazard use requiring only frontal protection.
**See appendix chart "Selection of Shade Numbers for Welding Filters."

APPLICATIONS		
OPERATION	**HAZARDS**	**RECOMMENDED PROTECTORS:** Bold Type Numbers Signify Preferred Protection
ACETYLENE—BURNING ACETYLENE—CUTTING ACETYLENE—WELDING	SPARKS, HARMFUL RAYS, MOLTEN METAL, FLYING PARTICLES	7, 8, 9
CHEMICAL HANDLING	SPLASH, ACID BURNS, FUMES	**2**, 10 (For severe exposure add 10 over 2)
CHIPPING	FLYING PARTICLES	**1, 3, 4, 5, 6, 7A, 8A**
ELECTRIC (ARC) WELDING	SPARKS, INTENSE RAYS, MOLTEN METAL	**9, 11** (11 in combination with 4, 5, 6, in tinted lenses, advisable)
FURNACE OPERATIONS	GLARE, HEAT, MOLTEN METAL	**7, 8, 9** (For severe exposure add 10)
GRINDING—LIGHT	FLYING PARTICLES	**1, 3, 4, 5, 6**, 10
GRINDING—HEAVY	FLYING PARTICLES	**1, 3**, 7A, 8A (For severe exposure add 10)
LABORATORY	CHEMICAL SPLASH, GLASS BREAKAGE	**2** (10 when in combination with 4, 5, 6)
MACHINING	FLYING PARTICLES	**1, 3, 4, 5, 6**, 10
MOLTEN METALS	HEAT, GLARE, SPARKS, SPLASH	**7, 8** (10 in combination with 4, 5, 6, in tinted lenses)
SPOT WELDING	FLYING PARTICLES, SPARKS	**1, 3, 4, 5, 6**, 10

Courtesy of the American National Standards Institute.

Figure 11-3.—Selection of eye and face protective equipment. Illustration and table are taken from American Standard Z87.1, *Practice for Occupational and Educational Eye and Face Protection.*

185

ments of the wearer

4. Be durable, easily cleanable, and shall be capable of being disinfected

Protectors must be kept clean and in good repair. Every protector shall be distinctly marked to identify the manufacturer.

When limitations or precautions are indicated by the manufacturer, they shall be transmitted to the user, and care taken to see that such limitations and precautions are strictly observed.

Design, construction, testing, and use of devices for eye and face protection shall be in accordance with American National Standard *Practice for Occupational and Educational Eye and Face Protection,* Z87.1.

Users of corrective lenses. Persons whose vision requires the use of corrective lenses in spectacles, and who are required by this standard to wear eye protection, shall wear goggles or spectacles of one of the following types:

1. Spectacles whose protective lenses provide optical correction

2. Goggles that can be worn over corrective spectacles without disturbing the adjustment of the spectacles

3. Goggles that incorporate corrective lenses by mounting them behind the protective lenses

Types of eye protection. Eye protection equipment is available in many types and styles. Eye protection can be furnished in heat- or chemically treated glass, plastic, wire screen, or light-filtering glass. Employers and/or

supervisor should be familiar with these various forms of eye protection and should know the best for the job. Figure 11-3 is a selection chart for eye and face protection with recommendations for particular hazards and operations. Any equipment selected must meet respective design and performance requirements.

Cover goggles. This type of goggles is frequently worn over ordinary spectacles, protecting both the wearer's eyes and his corrective lenses, because lenses that have not been heat or chemically treated are easily broken. Cover goggle protect the spectacles underneath against pitting as well as breaking.

Cover goggles include the cup type with heat-treated lenses and the wide-vision type with plastic lenses. Both types are used for protection during heavy grinding, machining, chipping, riveting, work with molten metals, and similar heavy operations. They offer the advantage of being wide enough to protect the eye socket and being able to distribute a blow over a wide area.

Tempered lenses. This type of lens is widely used in ordinary eye-corrective spectacles for street wear, but many do not meet the requirements of American National Standard Z87.1-1968.

A worker who normally wears prescription lenses may have his correction ground in heat-treated lenses. A refractionist (ophthalmologist or optometrist) should do the fitting and should know the job for which eye protection is to be worn.

186

Protective spectacles. A spectacle without side shields may be worn where it is unlikely that particles will fly toward the side of the face, although spectacles with side shields are recommended for all industrial use. Side shield spectacles should always be worn where there is additional side exposure. Frames must be rigid enough to hold lenses directly in front of the eyes. The nose bridge can be adjustable or universal. Frames should be fitted by a person trained for the job.

Goggles—various types.

CHEMICAL GOGGLES, with soft vinyl or rubber frames, protect eyes against splashes of corrosive chemicals and exposure to fine dust or mist. Lenses can be of heat-treated glass or acid-resistant plastic. For exposures involving chemical splashes, the goggles are equipped with baffled ventilators on the sides. Goggles must be ventilated also for vapor or gas exposures. Some types are made to fit over spectacles.

LEATHER MASK DUST GOGGLES should be worn by employees who work around noncorrosive dust, as in cement and flour mills. The goggles have heat-treated or filter lenses. Wire-screen ventilators around the eye cup provide air circulation.

MINER'S GOGGLES, for underground work and other locations where fogging is a serious problem, are made of a corrosion-resistant wire screen, coated a dull black to reduce reflection.

MELTER'S GOGGLES come in spectacle and cup types with cobalt blue glass in graded shades.

Lenses with color in the upper half and clear glass in the lower half are also available. Frames are of leather or plastic to protect the face from radiant heat.

WELDER'S GOGGLES with filter lenses are available for such operations as oxyacetylene welding, cutting, lead burning, and brazing. OSHA regulation 29 CFR 1910.252(e) provides a guide for the selection of the proper shade numbers. These recommendations may be varied to suit the individual's needs and operations.

CHIPPER'S GOGGLES, which have contour rigid plastic eyecups, come in two styles—one for individuals who do not wear spectacles, and one to fit over corrective spectacles. Chipping goggles should be used where maximum protection from flying particles is indicated.

Other considerations.

GOGGLE FRAMES. All goggle frames should be of corrosion-resistant material that will neither irritate nor discolor the skin, that can withstand sterilization, and that are flame resistant or nonflammable. Metal frames should not be worn around electrical equipment or near intense heat. To permit the widest range of vision, goggles should be fitted as close to the eyes as possible without the eyelashes touching the lens. The lenses should have no appreciable distortion or prism effect.

CLEANING GOGGLES. Goggles should be sterilized before they are reissued to other employees. The proper procedure is to disassemble and wash them with soap or detergent in warm water. Rinse thoroughly. Immerse all parts in a solution containing germicide, deodorant, and a fungicide. Do not rinse, but hang to dry in air. Then place the parts in a clean, dustproof container.

If a lens has more than just the most superficial scratch, nick, or pit, it should be replaced. Such damage can materially reduce protection afforded the wearer.

Laser beam protection. Although no one type of glass offers protection from all laser wavelengths, glass for protection against any single laser wavelength can be had on special order from eyewear manufacturers. Typically, the eyewear will have maximum attenuation at a specific laser wavelength—with protection falling off rather rapidly at other wavelengths.

Laser goggles designed for protection from specific laser wavelengths should not be mistakenly used with different wavelengths of laser radiation. To avoid such an occurrance, distinctively colored frames are recommended with the optical density shown on the filter. Technical details, uses, hazards, and exposure criteria for lasers are given in the National Safety Council's *Accident Prevention Manual for Industrial Operations* and *Fundamentals of Industrial Hygiene.*

Face protection

The general and regulatory requirements for face and eye protection are covered in an earlier section of this chapter. Many types of personal protective equipment shield the face (and sometimes the head and neck

also) against light impact, chemical or hot metal splashes, heat radiation, or other hazards. The employer should be familiar with the type best suited for his operation (see Figure 11-3, page 185).

Face shields. Clear plastic face shields protect eyes and face of a person who is sawing or buffing metal, sanding or light grinding, or handling chemicals. The shield should be slow burning and must be replaced when warped or scratched. Also, plastic shields tend to become brittle with age and require replacement. The headgear and shield should be adjustable to the size and contour of the head, and be easy to clean.

Metal screen face shields deflect heat from a person and still permit good visibility. They are used around blast furnaces, soaking pits, open hearth furnaces, and other sources of radiant heat.

Babbitting helmets. This type of helmet is used to protect the head and face against splashes of hot metal, rather than against heat radiation. The helmet consists of a window made of extremely fine wire screen, a tilting support, an adjustable headgear, and a crown protector. A chin protector can also be had if required.

Welding helmets. Welding helmets, shields, and goggles protect the eyes and face against the splashes of molten metal and the radiation produced by arc welding. The shell of the helmet must resist sparks, molten metal, and flying particles. It should be a poor heat conductor, and a nonconductor of electricity. Helmets that develop pinholes or cracks must be discarded.

Helmets should have the proper filter glass to keep ultraviolet and visible rays from harming the eyes. See OSHA regulations.

Impact goggles worn under the helmet protect the welder from flying particles when the helmet is raised. Eye protection of the spectacle type with side shields is recommended as minimum protection from flash from adjacent work or from popping scale of a fresh weld (see Figure 11-4).

Welder's helpers should also wear proper eye protection, i.e., welding goggles or helmet, while assisting in a welding operation or while chipping flux and scale from fresh welding.

Hand-held shields. A hand-held shield can be used where the convenience of a helmet is not needed, such as for inspection work, tack welding, and other operations requiring little or no welding by the user. Frame and lens construction is similar to that of the helmet. Welding goggles with filter glass shades up to No. 8 should be used in conjunction with hand-held shields. If darker shades are needed, then complete face protection is needed because of the danger of skin burns. When shades darker than No. 3 must be used, a side shield or cup goggles is recommended.

Acid-proof hoods. Acid-proof hoods that cover head, face, and neck are used by persons exposed to possible splashes from corrosive chemicals. This type of hood has a window of glass or plastic that is securely joined to the hood

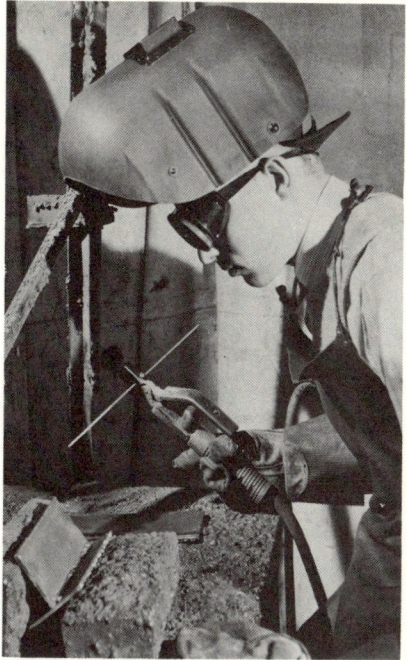

Figure 11-4.—Double eye protection for electric welders. Flash goggles protect welder's eyes from the arcs of other welders when he lifts his helmet to inspect the work.

to prevent acid from seeping through. Hoods made of rubber, neoprene, plastic film, or impregnated fabrics are available for resistance to different chemicals. Consult the manufacturer to find the protective properties of each material.

Air supplied hoods. Hoods with an air supply should be worn for work around toxic fumes, dusts, gases, or mists. Because an unventilated hood quickly becomes warm and humid, an air line also makes the hood more comfortable. The worker should wear a harness or

189

belt to support the hose.

Transparent face shields are used where there is limited exposure to direct splashes of corrosive chemicals. Splashproof cup goggles should be worn under the shield for added protection.

Ear protection

Need for hearing protective devices. Protection against the harmful effects of noise must be provided when the employee is exposed to sound levels that exceed the permissible levels established in OSHA's noise regulations. When noise exceeds the permissible levels, and such noise cannot be reduced through administrative or engineering controls, personal protective equipment (hearing protective devices) must be used to reduce the exposure to within the permissible levels.

While there is still some disagreement as to the maximum intensity of sound to which the ear can be subjected without damage to hearing, the OSHA standards must be used for compliance purposes. These state that the permissible noise exposure for an 8 hour period is 90 dB, measured on a weighted scale that corresponds to the response of the human ear. This level may be increased slightly as the duration of exposure decreases. (Review the discussion in Chapter 13 for more details on this subject. Also refer to NSC's *Industrial Noise and Hearing Conservation.*)

Types of ear protection.

Available protection. Commercially available earplugs, if properly fitted and used, generally reduce noise reaching the ear by 25 to 30 dB (decibels, a measure of the loudness of sound) in the higher (more harmful) frequencies. They should give ample protection against sound levels of 115 to 120 dB. Earmuffs of the better type can reduce noise an additional 10 to 15 dB, making them effective against sound levels of 130 to 135 dB. Combinations of earplugs and muffs give 3 to 5 dB more protection. In no case, however, is total attenuation (sound reduction) more than about 50 dB, because conduction of noise through the bones of the head becomes significant at this point.

Insert-type protection. This type is inserted into the ear canals and varies considerably both in design and material (Figure 11-5c). Materials used are pliable rubber, soft or medium plastic, wax, and wax-impregnated cotton. (*a*) Rubber and plastic types are popular because they are easy to keep clean, are inexpensive, and give good performance. (*b*) Wax tends to lose its effectiveness during the work day because jaw movement changes the shape of the ear canal and this breaks the acoustic seal between ear and insert. (*c*) Cotton alone is a poor choice because of its low attenuating properties. However, if paraffin wax is mixed into the cotton, it becomes much more efficient. Glass down, so-called "Swedish wool," is being used with success by some. Still another type is specially molded to each ear; because each person's ear canal is differ-

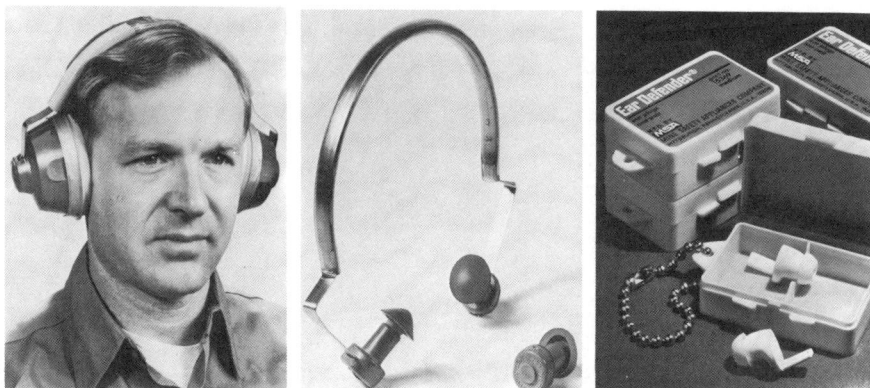

Figure 11-5.—Three major types of industrial hearing protective devices: (*a*) circumaural protectors (ear muffs), (*b*) supra-aural or semi-aural (ear caps), and (*c*) aural (ear plugs).

ent, these plugs become the property of the individual to whom they are fitted. These plugs must be fitted by a trained, qualified person.

Skin irritations, injured ear drums, or other harmful reactions are exceedingly rare when properly designed, well fitted, and clean hearing protectors are used. They should cause no more difficulty than does a pair of well fitted safety goggles.

Muff types. Cup or muff devices cover the external ear to provide an acoustic barrier, the effectiveness of which varies with the size, shape, seal material, shell mass, and type of suspension of the muff. Figure 11-5a illustrates one type of muff. Head size and shape also influence the effectiveness. Liquid- or grease-filled cushions give better noise suppression than plastic or foam rubber types. Muffs are made in a universal type or a head, neck or chin type. Professional assistance should be sought to make certain that the hearing protection is the best suited for the problem.

Hand and finger protection

Need for protection. Fingers and hands are exposed to cuts, scratches, bruises, and burns and are a major source of disabling injuries. Although fingers are hard to protect (because they are needed for practically all work), they can be shielded from many common injuries by use of proper protective equipment, such as that described in the following sections. (See Figure 11-6.)

Types of equipment available. ASBESTOS GLOVES protect against burns and discomfort when the hands are exposed to sustained conductive heat.

METAL MESH GLOVES, used by those who work constantly with knives, protect against cuts and blows from sharp or rough objects and tools. (See Figure 9-8, Chapter 9, page 151.)

RUBBER GLOVES, worn by electricians, should be tested regular-

191

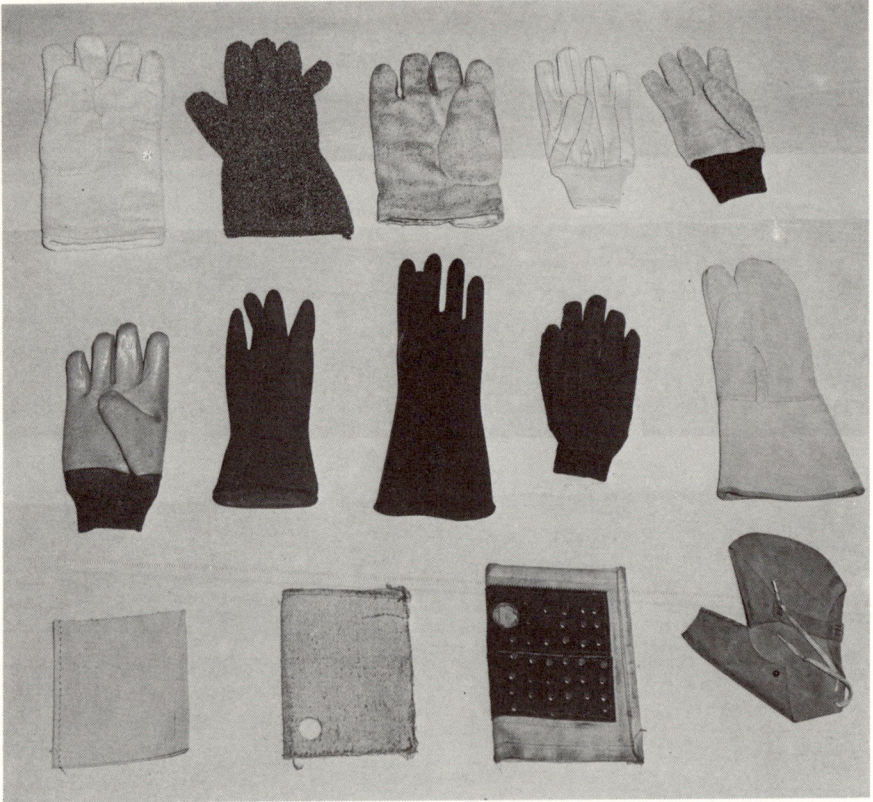

Figure 11-6.—Pictured are various types of specialized hand protectors. **In the top row (l. to r.)** are asbestos, loop pile, and aluminized gloves, all heat resistant; 19-ounce cotton glove and loop pile glove are for general shop wear. **In the middle row** are plastic dipped gloves for oily work, fine rubber gloves for protection against acids, etc., where finger dexterity is required, neoprene gauntlet for acid, caustic handling, neoprene and cork dipped glove for slippery and oily jobs, and a chrome leather welder's glove for arc welding. **In the bottom row** are a neoprene sandwich palm pad for protection against sharp edges, asbestos palm pad for hot sharp edges, brass studded palm pad for handling heavy material, and open-back leather palm pad for annealing operations.

ly for dielectric strength.

RUBBER, NEOPRENE, OR VINYL GLOVES are used when handling chemicals and corrosives. Neoprene and vinyl are particularly useful when petroleum products are handled.

LEATHER GLOVES resist sparks, moderate heat, chips, and rough objects. They provide some cush-ioning against blows. They are generally used for heavy work. Chrome-tanned leather or horsehide gloves are usually used by welders.

CHROME-TANNED COWHIDE LEATHER GLOVES with a steel-stapled leather patch or steel staples, on palms and fingers, are used in foundries and steel mills.

192

COTTON OR FABRIC GLOVES are suitable for protection against dirt, slivers, chafing, or abrasion. They are not heavy enough for use with very rough, sharp, or heavy materials.

COATED FABRIC GLOVES protect against moderately concentrated chemicals. They are recommended for use in canneries, packing houses, food handling and similar industries.

Specially made electrically tested rubber gloves, worn under leather gloves which prevent punctures, are used by linemen and electricians who work with high voltage equipment. A daily visual inspection and air test (by mouth) of the rubber gloves must be made. Rubber gloves should extend well above the wrist so that there is no gap between the coat or shirt sleeve and glove.

Employees usually do not mind wearing gloves when the danger of not wearing them is apparent. When a worker no longer has to worry about cutting his hands, and when he can grip materials better because of his gloves, production increases. Still, if employees become lax in wearing their gloves, stricter supervision is needed. The supervisor should be aware that gloves may be a hazard when worn around certain machining operations. Their use should not be allowed where such hazards exist.

HAND LEATHERS. Hand pads or hand leathers are often more satisfactory than gloves for protecting against heat, abrasion, and splinters. Wristlets or arm protectors are available in the same material as gloves.

Protective creams. Another source of hand protection is found in the various protective creams, or "invisible gloves" as they are often called. They may be used to protect skin against many irritants when protective equipment is not practicable. Creams are available in water-soluble or water-resistant types, and in several grades for different exposures. Water-soluble creams protect against cutting oils, paints, lacquers, and varnishes. Water-resistant types are used where the cutting oil or cooling lubricant has more than 10 percent water content. Soap and warm water will remove creams of this type. To be effective, cream coatings should be renewed frequently. It should be remembered that creams do not protect against highly corrosive substances.

Foot and leg protection

Need for protection. In most instances, the severity of a disabling foot injury may have been significantly reduced if the injured person had been wearing appropriate foot protection. Foot protection is needed in most industries, and employers must see that their employees wear the proper protection. When employees are exposed to *foot* and *leg* injuries, the employer must see that appropriate protection is worn.

Types of foot protection.

Safety shoes and foot guards. Foot and toe protection is available in various types and styles, each for a specific purpose. (See

193

Figure 11-7.—Typical safety shoes. Second from left illustrates metal instep guards that fit over shoes. *Inset:* Shoe with built-in instep protection.

Figure 11-7.) The use of some safety shoes with metal toe boxes are prohibited where the danger of contact with hot electrical equipment may exist.

Available shoe styles and types. Some examples of typical safety shoes used for particular exposures are as follows:

METAL-FREE SHOES, boots, and other footwear are available for use where there are severe electrical hazards or fire and explosion hazards.

"CONGRESS" OR GAITER-TYPE SHOES are used to protect people from splashes of molten metal or from welding sparks. This type can be removed quickly to avoid

serious burns. These shoes have no laces or eyelets to catch molten metal.

REINFORCED SHOES having inner soles of flexible metal are designed for use where there are hazards from protruding nails and where the likelihood of contact with energized electrical equipment is remote, for example, in the construction industry.

FOR WET WORK CONDITIONS, like those found in dairies and breweries, leather shoes with wood soles or wood soled sandals are effective. Wood soles provide good protection on jobs that require walking on hot surfaces (that are not hot enough to char wood). Wood soles have been so

194

generally used by men handling hot asphalt that they are sometimes called "paver's sandals."

SAFETY SHOES with metatarsal guards should always be worn during operations where heavy materials, such as pig iron, heavy castings, and timbers, are handled. They are recommended any time there is a possibility of objects falling and striking the foot above the toe cap. Metal foot guards are long enough to protect the foot back to the ankle and may be made of heavy-gage, flanged, and corrugated sheet metal. They should be easy to adjust and remove.

SHOE DESIGN. It is important to remember that safety toe shoes are primarily designed to provide toe protection and will *not* offer total foot protection. The old complaint that safety shoes are too hot, too heavy, and too uncomfortable has been eliminated. New designs now make some safety shoes as comfortable, as practical, and as attractive as street shoes. The steel cap weighs about as much as a pair of rimless eyeglasses or a wrist watch. The toe box is insulated with felt to keep the toes from getting too hot or cold, and some shoes have inner soles of foam latex, which consist of tiny "breathing cells," which makes the wearing of safety shoes more comfortable.

Safety boots, shoes, and rubber overshoes. This type of equipment is available in several styles and types to fit a variety of specific needs. Safety boots, shoes, and rubber overshoes may be obtained with conductive soles to drain off static charges, and with nonferrous construction to reduce the possibility of friction sparks in locations with a fire or explosion hazard. Initial and subsequent periodic tests should be made on conductive footwear in order to make sure that the maximum allowable resistance of 450,000 ohms is not exceeded. Disposable shoe covers made of polyethylene are an example of specialized footwear designed for specific purpose, in this case to protect the product in a clean room.

Hygienic practices—foot protection.

Showering facilities. Where shower facilities are required for employees who are exposed to dangerous materials, many employers provide paper slippers or wooden sandals for each individual to reduce the possibility of foot infection. Paper slippers are discarded after a single use, while the sandals are disinfected at frequent intervals, especially before being assigned to others.

Reissuance of boots. Some employers find it necessary for men on different shifts or jobs to wear the same pair of rubber boots. Where such conditions are encountered, great care should be exercised to disinfect boots after each shift or job.

Leg protection.

Leggings, which encircle the leg from ankle to knee and have a flap at the bottom to protect the instep, protect the entire leg. The front part may be reinforced to give impact protection. Such guards are used by persons who work around molten metal. Leggings should permit rapid removal in case of emergency. Hard fiber or metal guards are available to protect shins against impact.

Knee pads protect employees whose work, like cement finishing or tile setting, requires much kneeling.

Selection and materials. Foot and leg protectors are available in many different materials, the type selected depends on the work being done. Where molten metals, sparks, and heat are the major hazards, for example, asbestos or leather is best. Where acid, alkalis, and hot water are encountered, natural or synthetic rubber or plastic, resistant to the specific exposure, may be used.

Other body protection

Knives and cleavers. Aprons are made of various materials, such as leather or fabric, with padding or stays, which offers protection against light impact and against sharp knives and cleavers, such as used in meat processing. (See Figure 9-8, Chapter 9.)

Machine shop aprons. An apron worn near moving machinery should fit snugly around the waist. Neck and waist straps either should be light strings or instant-release fasteners in case the garment is caught. Split aprons should be worn on jobs that require mobility on the part of the worker. Fasteners draw each section snugly around the legs.

Hot sparks and molten metal. Welders are often required to wear leather vests or capes and sleeves, especially when doing overhead welding, as protection against hot sparks and bits of molten metal. Flameproof gauntlet gloves and aprons are desirable. Woolen clothing is preferable to cotton. Trousers or overalls should not be rolled up, as this could catch sparks. Asbestos coats and aprons are often used by those who work around hot metal or other sources of intense conductive heat.

Chemicals, acids, and caustics. A rubber apron protects the body against the ordinary hazards of handling acids or caustic solutions in *small* quantities. Some liquids harm natural rubber, but synthetic rubber—especially neoprene—and some plastics give satisfactory resistance.

Whenever the danger of splash is present, precaution should be taken to see that all parts of the body which might be exposed are protected. Hands and arms should be protected by rubber gloves. A rubber or rubberized coat and hat should be worn for overhead hazards. Special precautions must be taken to decontaminate the clothing that has been used in corrosive atmospheres before it is used again. Before clothing is removed, it should be thoroughly washed with a hose stream.

Respiratory Protection
General requirements
Permissible practice. The general respiratory requirements in

196

this section have been drawn from OSHA regulations. Additional information may be obtained from the referenced regulations or from American National Standard Z88.2, *Practices for Respiratory Protection,* which was used by OSHA to promulgate these regulations.

In the control of those occupational diseases caused by breathing contaminated air, the primary objective is to prevent atmospheric contamination. This shall be accomplished as far as possible by accepted engineering control measures (for example, enclosure or confinement of the operation, general and local ventilation, and substitution of less toxic materials). When effective engineering controls are not feasible, or while they are being instituted, appropriate respirators shall be used.

Employer responsibility. Respirators must be provided by the employer when they are necessary to protect the health of the employee. The employer must provide the respirators which are applicable and suitable for the purpose intended. The employer is also responsible for the establishment and maintenance of a respiratory protective program which meets the minimum requirements.

Minimal program requiements. The employer is required to establish respiratory protection program activities which include, but are not limited to, written standard operating procedures governing the selection and use of respirators on the basis of hazards to which workers are exposed; the

instruction and training of the wearer in the proper use of respirators and their limitations; inspection, maintenance, cleaning, disinfecting and storage of equipment; appropriate surveillance of work area conditions, degree of employee exposure and suitability of respiratory protections used; and regular inspection and evaluation to determine the continued effectiveness of the program.

The employer should not assign persons to tasks requiring the use of respirators unless it has been determined that they are physically able to perform the work and use the equipment. The local physician shall determine what health and physical conditions are pertinent. The respirator user's medical status should be reviewed periodically (for instance, annually).

Employee responsibility. The employee shall use the provided respiratory protection in accordance with instructions and training received.

Equipment approvals

General requirements. Approved or accepted respirators shall be used when they are available. The respirator furnished shall provide adequate respiratory protection against the particular hazard for which it is designed in accordance with standards established by competent authorities.

Recognized authorities. The U.S. Department of Interior, Mining Enforcement Safety Administration, and the U.S. Department of Health, Education, and Welfare, National Institute of Occupational Safety and Health, are recognized as such authorities.

Although respirators listed by the U.S. Department of Agriculture continue to be acceptable for protection against specified pesticides, the U.S. Department of the Interior, Bureau of Mines, is the agency now responsible for testing and approving pesticide respirators.

Selection of respirators

Selection considerations. It is essential to have some knowledge concerning the various operational and environmental considerations that will influence the selection of the required protection. Among the many factors to be considered in the selection of the proper respiratory protection device for any given situation involving air contamination are the following:

1. The nature of the hazardous operation or process
2. The type of air contaminant, including its physical properties, its chemical properties, physiological effects on the body, and its concentration
3. The period of time for which respiratory protection must be provided
4. The location of the hazard area with respect to a source of uncontaminated respirable air
5. The functional and physical characterisitcs of respiratory protective devices

The general principles for selecting respiratory protective devices are outlined next. The discussion on health hazards in Chapter 13 may be helpful to those not too familiar with some of the terms used here.

Hazard considerations. The employer and/or the person assigned to the safety and health function should familiarize himself with the type of hazard for which a given type of respiratory equipment is approved, and he should not permit its use for protection against hazards for which it is not designed. For instance, particulate filter respirators are of no value as protection against solvent vapors, injurious gases, or lack of oxygen. Their use under these conditions is one of the most common and most dangerous abuses of respirators.

Other *common* misuses of respiratory equipment are to use chemical cartridge respirators where the nature of the hazard requires the use of gas masks, or to use chemical filtering types where atmosphere-supporting or self-contained units are necessary.

Some hazardous substances require protection against (*a*) damage to the respiratory system, and also (*b*) systemic injury through the skin. All substances should be investigated to determine whether this double hazard is involved. In addition, before respiratory equipment is ordered, it is best to discuss the type of exposure with knowledgeable people, i.e., safety specialists, industrial hygienists, and with manufacturers and dealers.

From the standpoint of fire hazard, neither pure oxygen nor air containing more than 21 percent oxygen is to be preferred to ordinary air for use in atmosphere (air) supplied respirators or self-contained breathing apparatus, for ventilating, or for other purposes. A flammable substance in the presence of oxygen requires only a small fraction of the energy to ignite that the same substance requires in the presence of air, and the ensuing fire or explosion is much more violent.

Air or oxygen provided by or supplied to any respiratory equipment must be free from contaminants; quality as well as quantity is important.

Classification of respirators. Respiratory protective devices can be classified as follows:

Air purifying respirators

Atmosphere (air) supplied respirators

Self-contained breathing devices

In turn, these are divided into a number of subclasses.

Air purifying respirators and where they are used

Gas masks. The gas mask type of air purifying respirator consists of a facepiece connected by a flexible tube to a canister (Figure 11-8). Contaminated air is purified by chemicals in the canister.

Because no one chemical has been found that will remove all gaseous contaminants, the canister (purifying agent) must be carefully chosen to fit the specific need. A canister designed for a specific gas or vapor (or a single class or gases or vapors) will give longer protection than a same-size canister designed for protection against a multitude of gases and

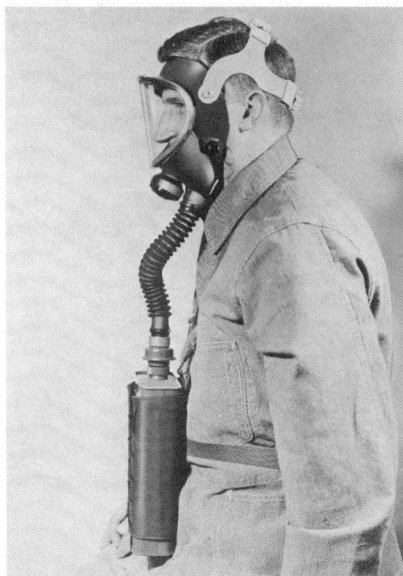

Figure 11-8.—A specific vapor canister gas mask.

vapors.

Canister gas masks with full facepiece are for emergency protection in atmospheres immediately dangerous to life. They *do not* provide protection against oxygen deficiency. In other words, contaminated air is purified; pure air is not supplied. The effectiveness of the canister type is limited to use in atmospheres containing at least 16 percent (by volume) oxygen, and not more than 2 percent of those toxic gases for which it is designed (except for ammonia, for which the limit is 3 percent, and phosphine, for which the limit is 0.5 percent). The period of protection that a gas mask provides depends upon (a) the type of canister, (b) the concentration of the gas or vapor, and (c) the activity of the user. Each per-

199

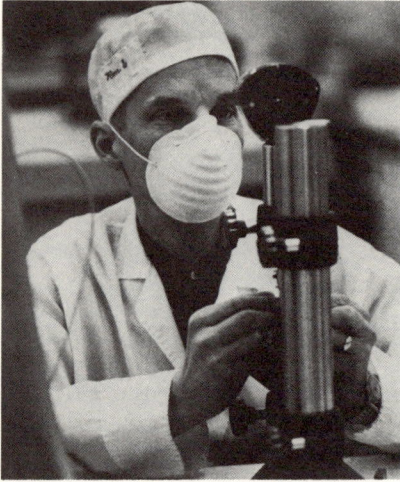

Figure 11-9.—Example of a nontoxic dust and particulate matter filter.

son who must use a gas mask should first undergo a physical examination, especially of his heart and lungs.

When a respirator is used in a gas or vapor that has little or no warning properties, like carbon monoxide, the canister must have an indicator or timer which shows when it should be changed for protection against the gas. Fresh canisters should be used each time a person enters the toxic atmosphere.

Chemical cartridge respirators. Chemical cartridge respirators consist of a half-mask facepiece connected directly to one or two small canisters of chemicals.

Chemical cartridge respirators are for use only in nonemergency situations; that is, for atmospheres which are harmful only after prolonged or repeated exposures.

Particulate filter respirators. A particulate (mechanical) filter respirator can be designed to give satisfactory protection against any kind of particle. (See Figure 11-9.) The major items to be considered are the resistance to breathing offered by the filtering element, the adaptation of the facepiece to faces of various sizes and shapes, use of safety glasses and earmuffs if necessary, the fineness of the particles to be filtered out, and their relative toxicity.

Combination respirators. Combination chemical and mechanical filter respirators utilize dust, mist, or fume filters with a chemical cartridge for dual or multiple exposure. The combination respirator is well suited for spray painting and welding.

Atmosphere (air) supplied respirators and where used

Hose masks. Hose masks are available with blower (power driven or hand operated), without a blower, or connected to a source of respirable air under pressure (air line hose mask, Figure 11-10).

Of the three, only the hose mask with blower is to be used for work in atmospheres where the hazard is immediate. The hazard is considered immediate if, in the event of failure of the equipment, hasty escape from the dangerous (toxic, flammable, and/or oxygen deficient) atmosphere would be impossible or could not be made without serious injury. In case of failure of the air supply on a hose mask with blower, it is still possible to breathe through the hose while making an escape.

Figure 11-10.—Constant-flow air line hose mask.

Hose masks are used by men entering tanks or pits where there may be dangerous concentrations of dust, mist, vapor, or gas, or insufficient oxygen. No one should enter, much less remain in a tank (or similar space) that tests show has less than 19.5 percent oxygen at any time, unless approved respiratory protective equipment is worn, such as a hose mask or self-contained breathing appartus. The atmosphere of the enclosed space should be tested to determine if there are any toxic or flammable contaminants present in dangerous or explosive concentrations. If so, or if oxygen is deficient, the space should be ventilated before anyone enters. The enclosure should be retested, as long as someone must work in it, and reventilated if necessary.

A hose mask with blower or a self-contained breathing apparatus must be used where, in the event of failure or the equipment, hasty escape from the dangerous atmosphere would be impossible or would result in serious injury. With either type, the user must wear a safety harness with lifeline attached and must be constantly attended by another man similarly equipped. In case of failure of the air supply in a hose mask with a blower, the wearer can still breathe through the hose while making his escape. (See Figure 11-11.)

Air line respirators. The air line respirator can be used in atmospheres not immediately dangerous to life, especially where working conditions demand continuous use of a respirator. Each person should be assigned his own respirator. Air line respirators are connected to a compressed air line. A trap and filter must be installed in the compressed air line ahead of the masks to separate oil, water, grit, scale, or other matter from the air stream. When line pressures are over 25 psig, a pressure regulator is required. A pressure release valve, set to operate if the regulator fails, should be installed. The air delivered to the mask should be of comfortable temperature and humidity.

To get clean air, keep the compressor intake away from any source of contamination, such as internal combustion engine exhaust. The compressor should have a temperature regulator or the compressed air line should have a carbon monoxide alarm in

201

Figure 11-11.—Man entering tank is equipped with lifeline held by standby man, similarly equipped. In addition, both are equipped with emergency egress unit (small cylinder on belt) in case air supply is accidently disconnected or stopped.

order to guard against the carbon monoxide hazard from overheated lubricating oil or from engine exhaust. The most desirable air supply is provided by a nonlubricated or externally lubricated medium-pressure blower, such as a rotary compressor. If a person must move from place to place, the air hose may be a nuisance. The supervisor must realize that such conditions will reduce a person's efficiency. Care must be exercised to prevent damage to the hose. For example, the hose should not be permitted to lie in oil.

Abrasive blasting respirators. Abrasive blasting respirators are

used to protect personnel engaged in shot, sand, or other abrasive blasting operations, which involve air contaminated with high concentrations of rapidly moving abrasive particles.

The requirements for abrasive blasting respirators are the same as those for an air line respirator of the continuous flow type, with the addition that mechanical protection from the abrasive particles is needed for the head and neck.

There are two types of abrasive blasting respirators that cover the head and neck and even the shoulder and chest. One type has a full facepiece (made of rubber) and an eyepiece (made of impact-resistant safety glass or

202

plastic covered by a metal screen). This unit is attached securely to a hood and cape made of tough, flexible rubber or rubber-covered fabric. Air is supplied to the full facepiece and is exhausted through an exhalation valve located under the hood in order to expell the air at the edges of the hood. The other type uses a rigid helmet (generally made of metal) to encase the user's head. The wearer is able to see by means of an impact-resistant safety glass or of a plastic covered by metal screen. An adjustable knitted fabric collar, covered with rubber- or plastic-coated fabric, fits over the metal helmet and down over the user's neck, shoulders, and chest in order to give additional protection. A flexible tube brings air to the helmet; generally a circular, perforated tube located near the top to the helmet is used for air distribution. Air, exhausted from the helmet, flows between the collar and the user's neck.

Air supplied hoods. For some long-term operations where a completely enclosed suit is not necessary, an air supplied hood may be used. These are particularly useful in hot, dusty stiuations.

Respirable air under suitable pressure should be delivered to a hood at a volume of at least 6 cfm.

Self-contained breathing devices

Use of devices. When people must work in an atmosphere immediately hazardous to life at distances from the source of fresh air, a self-contained breathing appartus, which carries its own sup-

Figure 11-12.—a 30-minute self-contained unit for entry into and escape from irrespirable atmospheres.

ply of oxygen, should be used. In an environment that contains a substance which is dangerously irritant or corrosive to the skin, a self-contained breathing device must be supplemented by impervious clothing.

Such devices afford complete respiratory protection in any toxic or oxygen-deficient atmosphere, regardless of the concentration of the contaminant. This equipment is frequently used in mine rescue work and in fire fighting. They also allow relative freedom of movement. (See Figure 11-12.)

Self-contained breathing apparatus should be worn only by men who are physically fit and well trained. Men should be refresher trained at least every six months in order to maintain efficiency.

Because of the extreme hazard, no one wearing self-contained

203

breathing apparatus should work in an irrespirable atmosphere unless other persons similarly equipped are standing by, ready to give help.

Types of devices. The three principal types of self-contained breathing apparatus are the compressed oxygen rebreathing type, the compressed oxygen (or air) nonbreathing type, and the oxygen-generating type. The length of time these units may be used is strictly limited. See National Safety Council, *Respiratory Protective Equipment,* Industrial Data Sheet 444, and the *Accident Prevention Manual.*

Use and care of respiratory equipment

Use of respirators. OSHA regulations require that standard procedures be developed for use of respiratory protection. These procedures should include all information and guidance necessary for their selection, safe use, limitation, and care. Possible emergency and routine uses of respirators should be anticipated and planned for. The type of respiratory protection for each job must be specified by a qualified individual supervising the program. In addition, written procedures must be prepared covering the safe use of respirators in dangerous atmospheres that may be encountered in normal operations or in emergencies. For safe use of any respiratory equipment the user must be trained by a competent person in all aspects of use, fitting, care, and limitations.

Details on the use of respiratory protection are in the OSHA standards as well as American National Standard Z88.2, the source document used by OSHA in promulgating these rules.

Maintenance and care of respirators.

Program requirements. A program for maintenance and care of respirators shall be adjusted to the type of plant, working conditions, and hazards involved, and shall include the following basic services: (*a*) inspection for defects (including a leak check), (*b*) cleaning and disinfecting, (*c*) repair, and (*d*) storage. Equipment shall be properly maintained to retain its original effectiveness.

Inspection. All respirators shall be inspected routinely before and after each use. A respirator that is not routinely used but is kept ready for emergency use shall be inspected after each use and at least monthly to assure that it is in satisfactory working condition. Air and oxygen cylinders shall be fully charged according to the manufacturer's instructions. The regulator and warning devices must be checked to ensure proper functioning. Respirator inspection shall include a check of the tightness of connections and the condition of the facepiece, headbands, valves, connecting tube, and canisters. A record shall be kept of inspection dates and findings for respirators maintained for emergency use.

Cleaning of respirators. Routinely used respirators shall be collected, cleaned, and disinfected as frequently as necessary to ensure that proper protection is provided for the wearer. Respirators maintained for emergency use shall be cleaned and disinfected after each use.

Storage of respirators. After inspection, cleaning, and necessary repair, respirators shall be stored to protect against dust, sunlight, heat, extreme cold, excessive moisture, or damaging chemicals. Respirators placed at stations and work areas for emergency use should be quickly accessible at all times and should be stored in compartments built for the purpose. The compartments should be clearly marked.

Safety Belts, Harnesses, and Lifelines

Safety belts and harnesses

General. Safety belts and harnesses with lifelines attached should be worn by those who work at high levels or in closed spaces where the air supply may not be adequate, and by those who work where they may be buried by loose material or be injured in confined spaces. This discussion does not include seat (vehicular safety) belts or linemen's belts.

Both normal and emergency types are available. Normal use involves comparatively light stresses applied during regular work—stresses that rarely exceed the static weight of the user. Emergency use means stopping a man when he falls—every part of the belt may be subjected to an impact loading many times the weight of the wearer.

A window cleaner's belt, for example, is subject to a moderate load most of the time it is used. It is subjected to a severe loading, however, if the man falls and only one terminal of the belt is attached. A belt for a person who leans back as he works should, therefore, have two D-rings, one on each side of the belt, to which a throw rope or lanyard can be attached. The rope then is connected to an anchorage.

A harness-type safety belt distributes the shock of an arrested fall. The harness permits a fallen person to be lifted with his back straight, rather than bent over a waist strap. This makes rescue easier if the victim is unconscious, buried, or must be taken out through a manhole.

If long falls are possible, the harness should be designed to distribute the impact force over the legs and chest as well as the waist, and where possible, a shock absorber or decelerating device can be used. To prevent a long fall, the line should be tied off overhead and should be as short as movements of the worker will permit.

Inspection and care of belts and lifelines

Belts and lines should be checked each time they are to be used. At least once every three months, belts should be examined by a trained inspector.

They should be maintained according to manufacturer's instructions.

Belts. Leather belts must be carefully checked for cuts and scratches on either side. Any deep vertical cut (crosswise to the belt) justifies discarding the belt. Lengthwise cuts should be viewed critically, and unless they are small, the belt should be discarded. Web belts should be examined for worn and torn fibers. When a number of the outer fibers are worn or cut through, the belt should be discarded. Belt hardware should be examined and worn parts replaced. If the belt is riveted, each rivet should be carefully inspected for wear around the rivet hole.

Lifelines. The outer surface of manila rope should be examined for cuts and for worn or broken fibers. Rope should be discarded when it has become smaller in diameter or has acquired a smooth look. Inner fibers of manila rope should be examined for breaks, discoloration, and deterioration. If it shows any of these signs, the rope should be discarded.

A steel wire rope should be examined for broken strands, rust, and kinks that may weaken it. Ropes must be kept clean, dry, and rust free. They should be lubricated frequently, especially before use in acid atmospheres or before exposure to salt water. After such use, wire rope should be carefully cleaned and again coated with oil.

Special Clothing

Impervious clothing

General. This type of clothing is used for protection against dusts, vapors, moisture, and corrosive liquids. There are many types of impervious materials which are fabricated into clothing of all descriptions, depending on the hazards to be protected from. The types of clothing range from aprons and bibs of sheet plastic to garments which completely enclose the body from head to foot and contain their own air supply.

Materials used include natural rubber, synthetic rubber, neoprene, vinyl, polypropylene, and polyethylene films and fabrics coated with them. Natural rubber is not suited for use with oils, greases, and many organic solvents and chemicals. Make sure that the clothing selected will protect against the hazards involved. Special impervious type clothing is required to be worn by employees whose work exposes them to carcinogens.

Cold weather clothing. In recent years thermal insulating underwear has become popular among outdoor workers because of its light weight. Two types are generally available. One is a thermal knit cotton patterned after regular underwear; the other consists of quilted materials, such as polyester fiber quilted between nylon. While this material does not catch fire any easier than cotton, once it starts burning, the nylon and polyester melt, forming a hot plastic mess not unlike hot pitch, which can adhere to skin and cause serious burns. Quilted insulating underwear is now available that has been made fire retardant to combat this danger.

Other special clothing. Safety experts have been very ingenious in developing many highly specialized types of clothing for protection against special hazards. A partial list includes:

High visibility and night hazard clothing for construction, utility, maintenance workers, police, and firemen whose work exposes them to traffic hazards.

Disposable clothing made of plastic or reinforced paper is available for exposure to low level nuclear radiation and for use in the drug and electronic industries where contamination (to the product) may be a problem.

Leaded clothing of lead glass fiber cloth, leaded rubber, or leaded plastic for use by laboratory workers and other personnel exposed to x-rays or gamma radiation.

Electromagnetic radiation suit which provides protection from the harmful biological effects of electromagnetic radiation found in high level radar fields and similar hazardous areas.

Conductive clothing, made of a conductive cloth, is available for use by linemen doing bare-hand work on extra-high voltage conductors. Such clothing keeps the worker at the proper potential.

For special applications, manufacturers have a vast number of materials they can draw upon to meet specific hazards.

Policy For Care and Use of Personal Protective Equipment

Establishment of program policy

The employer, regardless of the size of his operation, must establish and post a clear-cut policy with respect to the company's personal protective program. It is essential that such policy be consistent with prevailing regulations as to availability, selection, proper use, inspection, care and maintenance of the equipment. Since the law makes the employer responsible for "employee-owned equipment," such equipment must of necessity receive the same attention as company-owned equipment to assure its adequacy, safety, use, and proper care and maintenance. A clearly defined policy assigning responsibility to supervisors to enforce the use of personal protective equipment is essential as is a rigid policy for employee compliance.

Employee training in use of equipment

Once an effective personal protective equipment program has been established, employees must be trained in the proper use, care, and limitation of the equipment. Further, it must be assured that such training is satisfactorily completed before the employee is permitted to operate the equipment.

Fire Protection

Basic Chemistry of Fire

Ordinary fire (one that can be combatted by ordinary extinguishing means) results from the combination of fuel, heat, and oxygen. When a substance that will burn is heated in air to a certain critical temperature called its "ignition temperature," it will ignite and continue to burn as long as there is *fuel,* the *proper temperature of the fuel,* and a *supply of oxygen.*

Recent studies in fire chemistry indicate that the original fuel molecules combine with oxygen in branched-chain reactions responsible for the evolution of flames. So, another factor, the *flame chain reaction,* must be considered.

Knowledge of the chemical reaction of a fire forms the basis for knowing how to prevent fire, and for knowing how to extinguish it.

Heat can be taken away by cooling—or can be kept away from the fuel source originally. Oxygen can be taken away by excluding the air—or by diluting the oxygen content. Fuel can be removed to an area where there is no flame—or protected from sources of ignition. And, the

chemical reaction can be stopped by inhibiting the rapid oxidation of the fuel.

Cooling

In order to extinguish a fire by cooling, it is only necessary to absorb a small portion of the total heat being evolved by the fire. The most common and practical agent is water, applied in the form of a solid stream, finely divided spray (fog), or incorporated in foam. Water also absorbs heat when it vaporizes—more than do other common extinguishing agents. Also, when vaporized into steam, it expands about 1700 times; this reduces the volume of air (oxygen supply) available to sustain combustion in the fire zone. Thus water is an efficient cooling medium and a diluting agent.

Removing fuel

Often, taking the fuel away from a fire is difficult and dangerous, but there are exceptions which should be considered in the design and engineering stages.

Flammable liquid storage tanks can be arranged so their contents can be pumped to an isolated empty tank in case of fire. If a flammable gas catches fire as it flows from a pipe, the fire will go out if the gas flow can be shut off. Also, in any mixture of fuel gases or vapors in air, adding an excess of air has the effect of diluting the fuel concentration below the minimum combustible concentration point. Therefore, an air blast may extinguish a fire if the vapor-air mixture is diluted below the lower flammable limit or if the flame is moved away from the fuel source at a velocity greater than the flame propagation rate.

Limiting oxygen

Extinguishment by separation of oxygen from fire can be accomplished through smothering, such as, by covering the burning area with a wet blanket (make sure the blanket is not made of highly flammable fibers), throwing dirt or sand on the fire, or covering it with chemical or mechanical foam. Extinguishment by diluting the reactants—oxygen and fuel vapor—below the concentration necessary to support combustion is accomplished in blanketing the fire area with carbon dioxide or noncombustible vaporizing liquids. The fire will remain out if the blanket is maintained long enough for the combustible material to cool below its self-ignition temperature and if no external ignition sources are present.

Carbon dioxide and vaporizing liquid are of limited value on fires involving wood, rags, or paper, because the blanket usually cannot be maintained long enough for all smoldering ignition

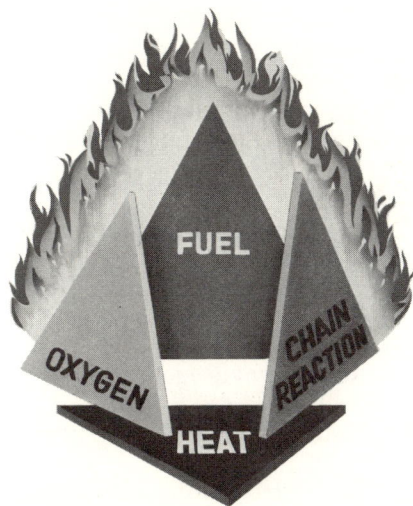

Figure 12-1.—The fire pyramid shows the four components necessary to produce ordinary burning. Remove any one and the fire goes out. Speed the process and an explosion results.

sources to be extinguished. Moreover, smothering is ineffective on materials that contain their own oxygen supply, such as ammonium nitrate or nitrocellulose.

In the field of fire prevention, the principle of separating oxygen from the fuel supply is also applied when an inert gas is used to purge operations involving flammable vapors, dusts, and other combustible materials under confined conditions when a source of ignition may exist.

Interrupting the flame chain reaction

In analyzing the anatomy of a fire, the original fuel molecules appear to combine with oxygen in

209

a series of successive intermediate stages, called branched-chain reactions, in arriving at the final end products of combustion. It is these intermediate stages which are responsible for the evolution of flames.

As molecules fragmentize in these branched-chain reactions, unstable intermediate products called *free radicals* are formed. The concentration of free radicals, such as hydrogen $(H-)$ and hydroxyl groups $(OH-)$, are the determining factors of flame speed.

It is the free radicals in these branched-chain reactions which are removed from their normal function as a chain carrier by dry chemical and halogenated hydrocarbon extinguishing agents.

Since each of these four elements of fire are interdependent, we can pictorialize fire extinguishment with a "fire pyramid," see Figure 12-1.

Classification of Fires

Four general classifications of fires have been adopted by the National Fire Prevention Association based upon the types of extinguishing media necessary to combat each. (See later discussion of portable fire extinguishers.)

Class A fires

Class A fires are those that occur in ordinary materials such as wood, paper, excelsior, rags, and rubbish. The quenching and cooling effects of water or of solutions containing large percentages of water are of first importance in extinguishing such fires. Special dry chemical agents (multi-purpose dry chemical) provide rapid knockdown of the flames and the formation of a coating that tends to retard further combustion.

Class B fires

Class B fires are those that occur in the vapor-air mixture over the surface of flammable liquids such as gasoline, oil, grease, paints, and thinners. The limiting of air (oxygen) or the combustion-inhibiting effect is of primary importance on fires of this class. Solid streams of water are likely to spread the fire (because the burning liquid floats on the water), but under certain circumstances water fog nozzles may prove effective. Generally, regular dry chemicals, multipurpose dry chemicals, carbon dioxide, foam, and halogenated hydrocarbon agents are used.

Class C fires

Fires that occur in or near electrical equipment where nonconducting extinguishing agents must be used are called Class C fires. Dry chemical, carbon dioxide, compressed gas, and vaporizing liquid extinguishing agents are suitable. Foam or a stream of water should *not* be used because both are good conductors and could expose the operator to a severe shock hazard.

Class D fires

Fires that occur in combustible metals (such as magnesium, titanium, zirconium, lithium, and sodium) are classified as Class D fires. Specialized techniques, extinguishing agents and extinguishing equipment have been developed to control and extinguish fires of this type. Normal extinguishing agents should generally not be used because there is a danger of increasing the intensity of the fire due to the chemical reaction between some extinguishing agents and the burning metal.

Degree of Fire and Other Hazards of Materials

Many chemicals have multiple hazards of varying degrees, such as flammability, toxicity, and reactivity (stability) hazards. Fires and other emergency situations often involve such chemicals and aggravate already existing problems due to the processing of such materials.

Information on these relative hazards must be readily available to those confronted with such emergencies if life safety, fire prevention, and effective fire control are to be achieved.

A system for the quick identification of hazardous properties of stored chemicals has been developed by the National Fire Protection Association. For uniformity, this system recommends the use of a diamond-shape symbol and numerals indicating the degree of hazard (see Figure 12-2).

The three categories of hazards are identified for each material: health, flammability, and reactivity (stability). The order of severity in each category is indicated numerically by five devisions ranging from 4, which indicates a severe hazard, to 0, which indicates that no special hazard is involved.

(See NFPA Standard 704M for standard ways of classifying the hazard degrees.)

Colors may also be used to better identify each category: *blue* is for health, *red* for flammability, and *yellow* for reactivity.

At the bottom of the three-part diamond is an open space which may be used for additional information such as radioactivity hazards, proper fire extinguishing agent, skin hazard, pressurized containers, protective equipment required, or unusual reactivity with water. The recommended signal to indicate an unusual reactivity with water and to alert the fire-fighting personnel *not* to use water is the letter "W" with a long line through the center.

Solvents

A solvent may be safe with respect to fire hazards but very unsafe with respect to health hazards. For example, carbon tetrachloride is nonflammable, but its vapors are extremely toxic. Before a solvent is used, therefore, it is necessary to determine both its toxic properties and its flammable properties.

A satisfactory solvent is inhibited methyl chloroform, or a blend of stoddard solvent and perchloroethylene. These solvents are commonly used in industry since they are relatively nonflammable and have a relatively high threshold limit value with respect to toxicity. (Chapter 13 of this handbook discusses industrial solvents, the health hazards associated with their use, and various control measures.)

Many of the cleaning solvents

211

BLUE	RED	YELLOW
IDENTIFICATION OF HEALTH HAZARD	**IDENTIFICATION OF FLAMMABILITY**	**IDENTIFICATION OF REACTIVITY**
Type of Possible Injury	**Susceptibility to Burning**	**Susceptibility to Release of Energy**
Signal	Signal	Signal
4 Materials which on very short exposure could cause death or major residual injury even though prompt medical treatment were given.	**4** Materials which will rapidly or completely vaporize at atmospheric pressure and normal ambient temperature, and which will burn.	**4** Materials which are readily capable of detonation or of explosive decomposition or reaction at normal temperatures and pressures.
3 Materials which on short exposure could cause serious temporary or residual injury even though prompt medical treatment were given.	**3** Liquids and solids that can be ignited under almost all ambient temperature conditions.	**3** Materials that are capable of detonation or explosive reaction but require a strong initiating source, or that must be heated under confinement before initiation, or react explosively with water.
2 Materials which on intense or continued exposure could cause temporary incapacitation or possible residual injury unless prompt medical treatment is given.	**2** Materials that must be moderately heated or exposed to relatively high ambient temperatures before ignition can occur.	**2** Materials that are normally unstable and readily undergo violent chemical changes but do not detonate; also materials that may react with water violently, or that may form potentially explosive mixtures with water.
1 Materials which on exposure would cause irritation but only minor residual injury even if no treatment is given.	**1** Materials that must be preheated before ignition can occur.	**1** Materials that are normally stable, but that can become unstable at elevated temperatures and pressures, or that may react with water with some release of energy, but *not* violently.
0 Materials which on exposure under fire conditions would offer no hazard beyond that of ordinary combustibles.	**0** Materials that will not burn.	**0** Materials that are normally stable even under fire explosive conditions, and that are not reactive with water.

FIRE

HEALTH SAFETY

Figure 12-2.—The varying degrees of flammability, toxicity, and reactivity hazards of chemicals in storage can be shown by this standard identification system.

encountered in industry are not single substances, but mixtures of different chemicals, usually marketed under nondescriptive trade names or code numbers. Currently, there are no absolutely safe cleaning solvents. Therefore, before any commercially available solvent is used, it is essential to know its chemical composition. Without such knowledge, the hazards cannot be evaluated nor the required safety controls be used.

Causes and Sources of Fire

To eliminate the causes of fire, it is important to know how and where fires start. The following discussion of causes of fire is based upon an analysis of more than 25,000 fires reported to the Factory Mutual System. By reviewing these basic causes and by making a survey (discussed later), an employer can develop a fire prevention program personalized to his own operations.

Electrical equipment

Electric motors, switches, lights, and other electrical equipment exposed to flammable vapors, dusts, gases, or fibers present special problems. The NFPA's "National Fire Codes" designate the standard governing a particular hazard and indicate the special protective equipment needed. The NFPA standards classify certain locations as particularly hazardous:

• Class I Locations are those in which flammable gases or vapors are or may be present in quantities sufficient to produce explosive or ignitable mixtures.

• Class II Locations are those locations which are hazardous because of the presence of combustible dust.

• Class III Locations are those which are hazardous because of the presence of easily ignitable fibers or flyings, but in which such fibers or flyings are not likely to be in suspension in air in quantities sufficient to produce ignitable mixtures.

All components and utilization equipment used in a hazardous location shall be chosen from among those listed by a nationally recognized testing laboratory such as Underwriters Laboratories Inc. or Factory Mutual System (except custom-made components and utilization equipment).

Equipment approved for a specific hazardous location shall not be installed or intermixed with equipment approved for another specific hazardous location.

The employer shall make sure that all wiring components and utilization equipment are maintained as vapor, dust, or fiber tight as contemplated by their approvals. There shall be no loose or missing screws, gaskets, threaded connections, or other impairments to this tight condition.

Electrical fire hazards can exist in other than the designated hazardous locations. Temporary lights should be protected. Flexible cable and cords must be maintained in a safe condition.

Haphazard wiring, poor connections, and "temporary" repairs must be brought up to standard. Fuses should be the proper type and size. Circuit breakers should

213

be checked to see that they have not been blocked in the closed position, which results in overloading, and to see that moving parts do not stick.

Cleaning electric equipment with solvents can be hazardous since many solvents available for this purpose are both flammable and toxic. It is, of course, of utmost importance to use the safest cleaning solvents available.

Friction

Overheated transmission bearings and shafting where dust and lint accumulate, as in grain elevators, cereal and textile mills, and plastic, woodworking, and metalworking plants, are frequent sources of ignition. Bearings should be kept lubricated so that they do not run hot, and accumulations of flammable dust on housings should be removed as part of the housekeeping routine. Pressure lubrication fittings should be kept in place and oil holes of bearings should be kept covered to prevent combustible dust and grit from entering the bearings and causing overheating.

Special fire hazard materials

Certain materials must be kept isolated to prevent fire. For example, some chemicals, like sodium and potassium, decompose violently in the presence of water—they evolve hydrogen and ignite spontaneously.

Yellow phosphorus may ignite spontaneously on exposure to air. Other combinations, too numerous to mention here, may react with the evolution of heat and produce fire or explosion—in some cases, without air or oxygen being present. Such materials must be handled in a special manner. See National Fire Protection Association Standard No. 49, *Hazardous Chemicals Data.*

Also, some materials, principally the ethers, during long periods of storage may become unstable and eventually explosive. In such cases, using the oldest stock first contributes to both fire safety and good housekeeping—and seeing that this principle is followed is part of the supervisor's job.

Whatever materials are used or stored in an operation, it is important to know whether they explode when heated, react with water, heat spontaneously, yield hazardous decomposition products, or otherwise react in combination with other materials. (See Degree of Fire and Other Hazards of Materials. pages 211-213.)

The employer's best course of action is to obtain detailed information and be guided accordingly. The company's fire insurance carrier can be asked for help. The manufacturer should be required to provide the necessary information before a material is brought on the premises.

Welding and cutting

If at all possible, welding and cutting operations should be done in a separate, well-ventilated room with a fire-resistant floor. This safety measure is not, of course, always practical. Details were discussed in Chapter 9. Hot work permits are discussed later in this chapter and also in Chapter 9.

Open flames

There must be no open flames in or near spray rooms or spray booths. Occasionally, it may be necessary to do indoor spray painting or spray cleaning outside of a standard spray room or booth, which is provided with electrical equipment that is listed for the hazardous location, and a proper ventilating system. In such cases, adequate ventilation must be provided and possible ignition sources, such as spark-producing devices and open flames, must be eliminated.

Welding or blow torches should be placed and used so that their flames are at least 18 inches from wood surfaces. They should not be used in the presence of dusts or vapors, or near flammable liquids, paper, excelsior, or similar material. Torches should not be left unattended while they are burning.

Portable heaters

Gasoline furnaces, portable heaters, and salamanders present a serious fire hazard. Their use should be discouraged as much as possible.

Fuel used in salamanders and other portable indoor heaters should be restricted to liquefied petroleum gas, fuel oil, or kerosene. The area in which they are burned must be well ventilated since, like other heaters, they produce carbon monoxide.

All these heating devices require attention with respect to clearances and mounting. A clearance of 2 feet horizontally and 6 feet vertically should be maintained between a heater and any combustible material.

Flammable material overhead should be removed or shielded by noncombustible insulating board or sheet metal with an air space between it and the combustible material. A natural-draft hood and flue of noncombustible material should be installed.

As a fire precaution, each unit must be carefully watched. Fuel oil and kerosene salamanders should be shut down before they are moved or refueled. They should be allowed to cool before being refueled.

All portable heating devices should be equipped with suitable handles for safe and easy carrying. They also should be secured or protected against tipping and upsetting.

Of the portable heating devices mentioned, the salamander presents the most serious fire hazard if it is not properly handled. Salamanders are too often improvised of old steel drums or empty paint containers, with scrap wood, tar paper, or other waste used as fuel. This practice should be discouraged since it usually permits little control of sparks and smoke.

Hot surfaces

If possible, smoke pipes from heating appliances should not pass through ceilings or floors. If a smoke pipe must be run through a combustible wall, a galvanized doublethimble with clearance equal to the diameter of the pipe and ventilated on both sides of the wall must be provided.

Soldering irons, electric lamps, hot process metal, and related hot surfaces should be protected.

215

Smoking and metals

Management usually has a specific policy with regard to smoking by employees. "Smoking" areas as well as "no smoking" areas must be clearly defined and marked off with conspicuous signs. Reasons for these restrictions must be clearly explained to the employees, and rigid enforcement must be maintained all the time with no exceptions.

Fire-safe, metal butt containers should be provided in places where smoking is permitted. If the carrying of matches if prohibited, special lighter equipment should be kept in service in smoking areas.

"No Smoking" areas, especially when they include stairways and other out-of-the-way places, should be watched for evidence of discarded smoking materials.

Spontaneous ignition

Spontaneous ignition is a chemical action in which there is a slow generation of heat from the oxidation of a fuel until the ignition temperature of the fuel is reached. The fuel then begins to burn. Conditions leading to spontaneous ignition exist where there is sufficient air for oxidation but not enough ventilation to carry away the heat as fast as it is generated. Any factor that accelerates the oxidation while other conditions remain constant obviously increases the likelihood of such ignition.

Rags and waste saturated with linseed oil or paint often cause fires because no provision is made for the generated heat to escape. By keeping such refuse in air-

Figure 12-3.—Air-tight waste can is made of sheet metal, has self-closing cover.

tight metal containers with self-closing covers, the oxygen supply is limited and a fire will quickly extinguish itself. The containers should be emptied daily; see Figure 12-3.

The best precaution against spontaneous ignition is either total exclusion of air or good ventilation. The former is practicable with small quantities of material through the use of airtight containers. Ventilation can best be assured by storing materials in small piles or by "turning" a large pile at regular intervals.

To determine the progress of spontaneous heating, temperatures should be taken in the interior of the mass of material. Various locations within the mass should be checked. Exterior temperatures are not likely to provide a good index to the progress of the heating.

216

Static electricity

Sparks due to static electricity may be a hazard wherever there are flammable vapors or gases or combustible dusts. Precautions against static electricity are required in such areas. Static charges result from friction between small particles or from the contact and separation of two unlike substances, one or both of which are nonconductive.

Static charges can be produced in many ways. An example is the blue continuous spark which can be seen when friction tape is unrolled in the dark. Static charges may also be produced by the flow of flammable liquids through a nonconductive hose or by the passing of dry and powdered materials down a nonconductive chute or through a machine.

It is impossible to prevent the generation of static electricity under these circumstances, but the hazard of static sparks can be avoided by preventing the accumulation of static charges. Any one or more of the following methods may be used:

• Grounding

• Bonding

• Maintaining the relative humidity at a predetermined level

• Ionization of the atmosphere

A combination of these methods may be advisable in some instances where the accumulation of static charges presents a severe hazard.

Grounding is accomplished by mechanically connecting a conductive machine or vessel (in

Figure 12-4.—Bonding equalizes the potential between objects so that a spark will not occur between them.

which the generation of static may be a hazard) to ground by means of a conductor of low resistance. Bonding eliminates the difference in electrical potential between objects. (See Figure 12-4.)

An example of this difference can be seen when the attendant puts gasoline in a vehicle. He bonds the car to the fuel tank, which in turn is grounded (in the ground). He does this by first touching the metal end of the fuel hose to the vehicle fuel line; a bonding wire runs through the rubber hose to the fuel pump, which is connected to the tank.

Another method of grounding is to make the entire floor and structure of the building conductive so that all equipment in contact with it will be grounded. When the former method is used,

217

the supervisor must check the continuity of the ground circuit. With the latter method, it is important that the floor be free of wax, oil, or other insulating films.

When humidity is low, the hazard of static is greatest. When humidity is high, the moisture content of the air serves as a conductor to drain off static charges as they are formed. Where humidification is utilized to prevent the accumulation of static charges, the supervisor must make certain that an effective relative humidity—usually 60 to 70 percent—is maintained. However, the minimum humidity required for safety may vary over a considerable range under different conditions, and under some conditions the static charge cannot be controlled by humidification. Engineering authorities should be consulted.

When air is ionized, it has sufficient conductivity to prevent static accumulation. Ionization is produced by electrical discharges, radiation from radioactive substances, or gas flames. Only an electrostatic neutralizer designed for use in hazardous locations should be used; otherwise, the neutralizer may itself be a source of ignition of flammable vapor or dust. It must be kept in good condition.

Good housekeeping for fire safety

Good housekeeping is another important part of an effective fire protection program. It is imperative that the supervisor maintain strict discipline with his crew and enforce the housekeeping rules at all times. Each person should be held personally responsible for preventing the accumulation of unnecessary combustible materials in his work area.

Here are precautions to take:

• Combustible materials should be present in work areas only in quantities required for the job, and should be removed to a designated, safe storage area at the end of each work day.

• Quick-burning and flammable materials should be stored only in designated locations. Such locations always should be away from ignition sources and have special fire extinguishing provisions.

• Vessels or pipes containing flammable liquids or gases must have no leaks. Any spills should be cleaned up immediately.

• Workmen should be required to guard carefully against any part of their clothing becoming contaminated with flammable liquids. If contamination does occur, they must be required to change such clothing before continuing to work.

• Passageways and fire doors should be kept clear and unobstructed.

• Material must not obstruct sprinkler heads or be piled around fire extinguisher locations or sprinkler and standpipe controls. See the recommendations on page 172.

If there are no sprinklers, clearance of 3 feet between piled material and the ceiling is required to permit room for the use of hose streams. Double these distances when stock is piled more than 15 feet high. Be sure to check applicable codes.

Determining the Fire Hazards
To Be Controlled

To develop the best fire prevention and control program a supervisor or manager must (*a*) determine the specific fire problems existing in his area of operations and (*b*) plan and take action to solve them. He should take both steps with the help of the best technical advice he can get from experts in the fire department or otherwise.

He should devise an inspection checklist that names as many places, materials, procedures, classes of equipment, conditions, and circumstances possible— ones where fire hazards are likely to exist and which should be examined. See Figure 12-5. (Inspections were discussed in Chapter 4.)

On the list, specific fire-safe practices and installations should be cited concisely. Brief notations are made beside each of the described safe practices (or the boxes filled with a "yes" or "no") to indicate observance of them or the lack of it. The person making the inspection will be able to spot neglected precautions readily. The sample fire prevention checklist shown here can serve as a guide, but each employer or plant manager should make his own and include special points or factors and operations that are not listed in the broad example.

The NFPA *Inspection Manual,* a pocket-sized book, is valuable reference for the beginner as well as the experienced inspector in setting up and conducting fire inspections. Its content covers both common and special fire hazards, their elimination or safeguarding, building construction, and human safety in all types of properties.

A supervisor, when conducting an inspection within his department or of one particular job, may be amazed at the detailed points to be observed, some of which may have escaped his notice previously. Unless a person has had considerable experience in fire prevention, he should ask the company's fire insurance engineer to help him conduct inspections and to make recommendations for eliminating fire hazards. If company policy permits, the supervisor usually may ask the local fire department for help, too.

It is important that all fire inspections be made with a critical eye. Every shortcoming should be listed. A supervisor should not hedge on listing certain hazards with the thought that they might reflect poor supervision. Omission of a pertinent detail might result in a fire later.

Periodic inspections by supervisors are an important part of a company's fire protection program. However, the supervisor's responsibilities, under a complete program, extend further. As he becomes acquainted with actual or potential fire hazards, and after all physical corrections possible have been made, he should familiarize the men in his department with each hazard and explain how it relates to them individually.

If the man on the job understands the reason for the rule and the possible consequences if he does not follow it, he is more like-

219

FIRE PREVENTION CHECK LIST

ELECTRICAL EQUIPMENT

- [ ] No makeshift wiring
- [ ] Extension cords serviceable
- [ ] Motors and tools free of dirt and grease
- [ ] Lights clear of combustible materials
- [ ] Safest cleaning solvents used

- [ ] Fuse and control boxes clean and closed
- [ ] Circuits properly fused
- [ ] Equipment approved for use in hazardous areas (if required)
- [ ] Ground connections clean and tight

FRICTION

- [ ] Machinery properly lubricated
- [ ] Machinery properly adjusted and/or aligned

SPECIAL FIRE-HAZARD MATERIALS

- [ ] Storage of special flammables isolated
- [ ] Nonmetal stock free of tramp metal

WELDING AND CUTTING

- [ ] Area surveyed for fire safety
- [ ] Combustibles removed or covered
- [ ] Permit issued

OPEN FLAMES

- [ ] Kept away from spray rooms and booths
- [ ] Portable torches clear of flammable surfaces
- [ ] No gas leaks

PORTABLE HEATERS

- [ ] Set up with ample horizontal and overhead clearances
- [ ] Secured against tipping or upset
- [ ] Combustibles removed or covered
- [ ] Safely mounted on noncombustible surface
- [ ] Not used as rubbish burners

HOT SURFACES

- [ ] Hot pipes clear of combustible materials
- [ ] Ample clearance around boilers and furnaces
- [ ] Soldering irons kept off combustible surfaces
- [ ] Ashes in metal containers

SMOKING AND MATCHES

- [ ] "No smoking" and "smoking" areas clearly marked
- [ ] Butt containers available and serviceable
- [ ] No discarded smoking materials in prohibited areas

SPONTANEOUS IGNITION

- [ ] Flammable waste material in closed, metal containers
- [ ] Flammable waste material containers emptied frequently
- [ ] Piled material cool, dry, and well ventilated
- [ ] Trash receptacles emptied daily

STATIC ELECTRICITY

- [ ] Flammable liquid dispensing vessels grounded or bonded
- [ ] Moving machinery grounded
- [ ] Proper humidity maintained

HOUSEKEEPING

- [ ] No accumulations of rubbish
- [ ] Safe storage of flammables
- [ ] Passageways clear of obstacles
- [ ] Premises free of unnecessary combustible materials
- [ ] No leaks or drippings of flammables and floor free of spills
- [ ] Fire doors unblocked and operating freely with fusible links intact

EXTINGUISHING EQUIPMENT

- [ ] Proper type
- [ ] In proper location
- [ ] Unobstructed
- [ ] Clearly marked
- [ ] In working order
- [ ] Service date current
- [ ] Personnel trained in use of equipment

Figure 12-5.—Sample checklist serves as guide in drawing up inspection list for a work area. List should be reviewed regularly to keep it up-to-date.

220

ly to comply. Patient explanation and persistent enforcment, in every case, are two prime fire prevention duties of the supervisor.

Fire Extinguishment

Portable fire extinguishers

Principles of use. Even though a building may be equipped with automatic sprinklers or other means of fire protection, portable fire extinguishers should also be available and ready for emergency. "Portable" is applied to the manual equipment used on small fires.

To be effective, portable extinguishers must be:

• Reliable

• The right type for each class of fire that may occur in the area

• In sufficient quantity to protect against the exposure in the area.

• Located where they are readily accessible for immediate use

• Maintained in perfect operating condition, inspected regularly, checked against tampering, and recharged as required.

• Operable by area personnel who can find them and use them effectively and promptly

Classification of portable fire extinguishers. Portable extinguishers are classified to indicate their ability to handle specific classes and sizes of fires. This classification is necessary because of the constant development of improved and new extinguishing agents and devices, and because of the availability of larger portable extinguishers.

Labels on extinguishers indicate the class and relative size of fire that they can be expected to handle.

The following paragraphs are a guide to the selection of portable fire extinguishers for given exposures, in accordance with classifications set forth by the National Fire Protection Association.

Class A extinguishers for ordinary combustibles, such as wood, paper, and textiles, where quenching-cooling effect is required. The numeral indicates the relative fire extinguishing potential of each unit. For example, a 4-A unit can be expected to extinguish approximately twice as much fire as a 2-A unit. See Table 12-A.

Class B extinguishers for flammable liquid and gas fires, such as oil, gasoline, paint, and grease, where oxygen exclusion or flame interrupting effect is essential. The numeral indicates the area, in square feet, of deep-layer flammable liquid fire expected to be extinguished by an unskilled operator under emergency fire conditions. For example, a 10-B unit can be expected to extinguish 10 sq ft of flammable liquid fire.

Class C extinguishers for fires involving electrical wiring and equipment where the dielectric nonconductivity of the extinguishing agent is of first importance. For example, water-solution extinguishers must not be used on electrical fires because water conducts electricity and the operator may receive a shock from energized electrical equipment via the water. Class C extinguishers are not classified by a numeral

221

TABLE12-A
EXTINGUISHERS SUITABLE FOR CLASS A FIRES

Basic Minimum Extinguisher Rating for Area Specified	Maximum Travel Distances to Extinguishers	Areas To Be Protected per Extinguisher		
		Light Hazard Occupancy	Ordinary Hazard Occupancy	Extra Hazard Occupancy
1A	75 ft	3,000 sq ft	Not permitted Except as Specified*	Not permitted Except as Specified*
2A	75 ft	6,000 sq ft	3,000 sq ft	Not permitted Except as Specified*
3A	75 ft	9,000 sq ft	4,500 sq ft	3,000 sq ft
4A	75 ft	11,250 sq ft	6,000 sq ft	4,000 sq ft
6A	75 ft	11,250 sq ft	9,000 sq ft	6,000 sq ft
10A	75 ft	11,250 sq ft†	11,250 sq ft†	9,000 sq ft
20A	75 ft	11,250 sq ft†	11,250 sq ft†	11,250 sq ft†
40A	75 ft	11,250 sq ft†	11,250 sq ft†	11,250 sq ft†

*The protection requirements specified in this table may be fulfilled by several extinguishers of lower ratings for ordinary or extra hazard occupancies. Certain smaller extinguishers which are charged with multipurpose dry chemical are rated on Class B and Class C fires, but do not have sufficient effectiveness to earn a 1A rating; therefore, they shall not be used to meet these requirements.
†11,250 sq ft is considered a practical limit.

From National Fire Protection Association Standard No. 10, Table 3-2.1.

because a Class C fire is essentially either a Class A or Class B fire with the addition of energized electrical wiring and equipment. Therefore, the coverage of the extinguisher must be chosen on the basis of the burning fuel.

Class D extinguishers for fires in combustible metals, such as magnesium, potassium, powdered aluminum, zinc, sodium, tatanium, zirconium, and lithium. Persons working in areas where Class D fire hazards exist must be aware of the dangers in using Class A, B, or C extinguishers on a Class D fire, as well as the correct way to extinguish Class D fires. Class D extinguishers units are not classified by a numerical system and are intended for a special hazard protection only.

Marking of fire extinguishers. The following markings are used for extinguishers and/or extinguisher locations to indicate the suitability of the extinguisher for a particular class of fire. (See Figure 12-6). Extinguishers suitable

Extinguishers suitable for Class A fires should be identified by a triangle containing the letter "A." If colored, the triangle should be colored green.

Extinguishers suitable for Class C fires should be identified by a circle containing the Letter "C." If colored, the circle should be colored blue.

Extinguishers suitable for Class B fires should be identified by a square containing the letter "B." If colored, the square should be colored red.

Extinguishers suitable for fires involving metals should be identified by a five-pointed star containing the letter "D." If colored, the star should be colored yellow.

Figure 12-6.

for more than one class of fire may be identified by multiple symbols as described previously.

Where markings are applied to wall panels, in the vicinity of extinguishers, they should be of a size and form as to be easily legible at a distance of 25 feet.

Location of extinguishers. Extinguishers must be located close to likely hazards, but not so close that they would be damaged or cut off by a fire. They should be located along the normal path of egress. Where highly combustible material is stored in small rooms or enclosed spaces, the extinguisher should be located outside the door.

The location should be made conspicuous, see Figure 12-7, and must be kept readily accessible at all times.

Each extinguisher should have on it a data plate giving the class of fire for which it is intended, operating instructions, and servicing instructions. It should carry the label of either Underwriters Laboratories Inc., or Factory Mutual System to indicate that the unit has been approved.

The location and installation of portable fire extinguishers, fire pails, fire blankets, stretchers and other fire fighting equipment should be at strategic places about the shop or job-site.

Mounting heights are also spelled out by federal requirements. Extinguishers having a

Figure 12-7.—Fire equipment should be conspicuously located, appropriately marked, and inspected regularly. Top of extinguisher must not be more than 5 ft from the floor; if it weighs more than 40 pounds, 3½ ft.

223

Courtesy Paulsboro Laboratory, Mobil Research & Development Corp.

Figure 12-8.—Secretary is instructed in use of a pressure extinguisher. Session lasts $2\frac{1}{2}$ hours and includes home fire safety.

gross weight not exceeding 40 pounds shall be installed so that the top of the extinguisher is not more than 5 feet above the floor. Extinguishers having a gross weight greater than 40 pounds (except wheeled types) shall be installed so that the top of the extinguisher is not more than $3\frac{1}{2}$ feet above the floor.

The supervisor should make sure that each of his workers knows the location of the nearest unit and that each individual is impressed with the importance of cooperating to keep areas around extinguisher units clear.

It is advisable for the supervisor to have his people engage in drills in which they use extinguishers applicable to their particular work areas (Figure 12-8).

Every organization should have

a specific program for periodic inspection and servicing of portable fire extinguishing equipment. This routine work is usually outside the scope of the department foreman or supervisor. However, there are some things he can do to assist in this program. For example, during regular department inspections, he should double check each data card to determine the date when each extinguisher was last serviced. By doing so, he may catch an omission which might prevent a small blaze from becoming an inferno.

Nonportable extinguishing systems

Equipment used to extinguish and control fires is of two types—Fixed and portable. Fixed systems include water equipment (automatic sprinkler, hydrants,

224

standpipe hoses) and special pipe systems for dry chemical, carbon dioxide, and foam. Special pipe systems are applicable to areas of high fire potential where water may not be effective, such as storage tanks for flammable liquids, and on electric equipment.

Fixed systems, however, must be supplemented by portable fire equipment. These often can preclude the action of sprinkler systems because they can prevent a small fire from spreading as well as provide rapid extinguishment in the early stages of a fire. Fixed systems are briefly described here. More details are in National Fire Protection Association Standards.

Automatic sprinklers. Automatic sprinklers are the most extensively used installations of the fixed fire extinguishing systems. These systems are basic and have proved so effective that most fire protection engineers consider them the most important firefighting equipment. Nationwide figures prepared by the National Fire Protection Association indicate that sprinklers have an efficiency of 96.2 percent satisfactory performance.

There are six basic types of automatic sprinkler systems: wet pipe, dry pipe, preaction, deluge, combination dry pipe and preaction, and limited water supply.

Foam systems. Foam apparatus may be fixed or portable and may be automatically or manually operated. Discharge rates vary from 15 through 4000 gpm; the larger systems are employed to protect areas in which oil, paint, and asphalt are stored or used.

There are two types of foam—chemical and mechanical. Chemical foam is formed by a chemical reaction in which masses of bubbles of carbon dioxide, aided by a forming agent, produce a frothy foam. Mechanical foam is made by adding a liquid concentrate to water and mixing air into the solution.

Wetting agents. Wet water contains a chemical that reduces its surface tension, and thereby increases its penetrating, spreading, and emulsifying properties and, thus, its cooling ability. It can be applied as a liquid or as a foam.

Slippery water is formed when a polyethylene oxide polymer is mixed with water in small (0.003 percent) quantities. By reducing friction, slippery water increases the capacity of small-diameter hose.

Fog. Conventional sprinklers deliver a rather coarse spray, but with a special head a fog pattern can be developed. The more droplets water divides into, the more surface produced for a given amount of liquid water. Because a small amount of water can be made to vaporize into a large volume of steam, a fire can be extinguished by cooling and by smothering, with little water damage to the premises.

Carbon dioxide extinguishing systems. Fixed (local or floodtype) CO_2 systems are often installed for the protection of rooms that contain electrical equipment, flammable liquid or gas processes, dry cleaning machinery, and other areas where fire can be extinguished by diluting the oxygen content of the air or where water must not be used because of electrical hazard or the nature of the product.

Dry chemical systems. Dry chemical piped systems have been developed for situations where quick extinguishment is needed either in a confined area or for localized application, and where re-ignition is unlikely. The systems are adaptable to flammable liquid and electrical hazards, they are available for either manual or automatic operations, and can be activated at the system or by remote control. A rate-of-rise, heat-actuated device or an electrical release controls the automatic operation.

Installations can provide for simultaneous closing of fire doors, operating valves, windows, and ventilation ducts, as well as shutting off fans and machinery and actuating alarms.

Dry powder. Designed to extinguish burning metals, such as magnesium, alkali metals, or nuclear-reactor fuel metals, dry powder is usually dispensed by a special extinguisher or it can be simply applied with a shovel.

Halons. Halons, or halogenated hydrocarbons, resemble the higher molecular weight fluorocarbons used for refrigerants and plastics.

These compounds extinguish a fire by inhibiting the flame chain reaction and ion (free radical) formation. Halon 1301, chemically CF_3Br, is a liquefied gas. Only 3.3

percent by volume in air is required to put out a butane fire. Properly used, it has no harmful effects and if detection is early enough, it acts rapidly. Although quite expensive, it is about 10 times more efficient than CO_2, and it is used in areas where records or equipment cannot be allowed to suffer water damage.

Steam and inert gas. Steam jet systems are used to smother fires in small rooms or in closed containers such as heaters, drying kilns, smokehouses, asphalt-mixing tanks, and dry cleaning dryers. They require a large supply of steam and pose a personal injury hazard for burns.

Inert gas is used to replace a specified percentage of the oxygen in air. Carbon dioxide, nitrogen, helium, argon, or flue gas (mostly nitrogen, carbon dioxide, and water vapor) may be used. To be effective, the inert gas must reduce the amount of oxygen in the air from the normal 21 percent to between 2 and 16 percent, depending upon the type of combustible material involved and the type of inerting gas.

Other controls

Hot work permits. In an effort to establish some control over operations using open flames or producing sparks, many industrial firms have instituted hot work permit programs, which require that authorization be secured before equipment capable of igniting combustible materials is used outside areas normally specified. These were discussed in Chapter 9. Detailed information on hot work permit programs is contained in Industrial Data Sheet 522, "Hot Work Permits," available from the National Safety Council.

Flameproofing wood and fabrics.

Fire-retardant treatments. Wood cannot be made fire resistive, but it can be treated to *retard* both the rate of flame spread over its surface and the rate it contributes fuel to the fire. There are two methods for treating wood: impregnation and use of fire-retardant coatings. Water-soluble treatments are not suitable for exterior use.

Fire-retardant coatings reduce the hazard from small fires in existing structures. Only those coatings listed by Underwriters Laboratories Inc. should be used. Effective impregnation can usually be done only by companies equipped to do this work.

Flame-retardant fabrics. Fabrics cannot be made noncombustible, nor even resistant to charring or decomposition, but chemical treatment will reduce their flammability. Some treatments merely inhibit the rapid spread of flame, whereas others prevent flames and depress dangerous afterglow.

Where the only small sparks, small flames, or temperatures up to 400 to 500 F are anticipated, flame-retardant canvas is often preferred to asbestos or chrome-tanned leather particularly where flexibility, durability, strength, and resistance to abrasion are required. For more serious exposures, such as heavy welding, firefighting, or foundry work, as-

bestos or chrome leather is required.

Alarms, Emergency Plans, Employee Participation

A complete and fully effective fire program includes fire alarms, evacuation plans, and provisions for employee participation. However, despite the best laid preventive plans, a fire may occur. Whenever a fire occurs it is important to:

• Turn in the alarm right away, regardless of the size of the fire.

• Attempt to extinguish or control the fire with appropriate fire extinguishing equipment promptly, in order to minimize the hazard.

Employee participation

Regardless of the size of a company, some kind of fire fighting organization is advisable.

Companies or plants that employ a large number of people or have particularly hazardous operations should organize and train special fire brigades. Local fire departments and insurance companies can give valuable help. See NSC Data Sheet 588, *Fire Brigades.*

Companies with few employees can take advantage of their small size to train everyone in the techniques of firefighting peculiar to their own situation. If this is done there will always be trained people on hand. If only a few people are given this training, it is conceivable that at some time there will be no one around who has had the necessary training. All employees should be taught how to operate fire extinguishers. They should know where exits are and what their dities will be if a fire should develop. (See Figure 12-8, page 224.)

Alarm Systems

Plants in areas where municipal fire departments are available usually have a municipal alarm box close to the plant entrance or located in one of their buildings. Others may have auxiliary alarm boxes, connected to the municipal fire alarm system, at various points in the plant. Another system often used is a direct connection to the nearest fire station which may register by a water alarm on the sprinkler system or be set off manually. In many cases, the telephone may be the means for signaling a fire alarm.

Whatever the alarm system used, all employees should be given careful instruction in how, when and where, to report a fire. These three items are extremely important since many fires progress beyond control simply because someone did not know how or when or where to give the alarm.

Evacuation plans.

Prevention of fire is the primary objective of any fire protection program. Nevertheless, each program must include provisions to assure the safety of employees in the event of a fire. Essential among such provisions are those that will chart and facilitate the quick and orderly evacuation of personnel if a serious fire emergency occurs.

The supervisor can do much to prepare for the safety of employ-ees in his department in the event of a serious fire. As a matter of current routine, he should make sure that each man and woman knows the evacuation alarms, both the primary and alternate exits and escape routes, and what to do during and after an evacuation.

The supervisor should make sure that each person knows that he or she, upon being alerted, must proceed at a fast walking pace—not a run—to an assigned exit or, if it is blocked by smoke or fire, to the nearest clear one. Day in and day out, he should emphasize that exit routes through the department, the exists themselves, and fire doors must be kept unblocked. Furthermore, the supervisor should make it understood that possible damage and additional hazards can be avoided if each employee, upon a signal for evacuation, shuts down his or her machine before leaving it.

Such essential knowledge and habits to conform to emergency procedures will best be instilled in personnel by periodic fire drills. If company policy assigns the responsibility to the supervisor, he should plan and conduct practice evacuations at regular intervals within his area of operations, after availing himself of expert advice and technical assistance. He should also do everything possible to integrate fire emergency training in his department with plant-wide drills, and the overall evacuation plan. When it is impossible to hold drills, the supervisor should see that printed information is distributed periodically to employees.

228

Industrial Hygiene and Health

The extent of the employer's involvement with industrial hygiene or occupational health programs will depend largely upon the nature of his operations, the materials and equipment he uses, and the type of hazards his operations may generate. Health hazards in the work environment can, to a certain degree, be recognized by the employer once he gains sufficient knowledge about their charactertistics.

Endeavoring to find and apply solutions for the elimination and/or control of health hazards and harmful environmental exposures, the employer will require the assistance of one or several specialists. Medical examinations (placement, periodic, and termination) may be required for employees exposed to toxic and harmful materials. Audiometric testing, industrial hygiene surveys, environmental monitoring, and consultations if not available in-plant, can be obtained through outside sources and private professional consultants in both industrial hygiene and medical fields.

Above all, the employer must remember to keep records of all medical examinations, environmental tests, and employee exposure records. These must be meticulously maintained because the employer may be required to produce them at some future date. See Chapter 5, Accident Records and Reports.

Since there is no way of developing guidelines for an industrial hygiene and health program that will suit the needs of every employer, a general program has been outlined in this chapter. From it, the employer can obtain some idea of the program elements that may be applicable to his own operational needs and responsive to OSHA's compliance requirements.

Federal Regulations

Most every employer will be required to implement some element of an industrial hygiene and/or occupational health program in order to be responsive to the OSHAct and its health regulations.

Health standards

The Occupational Safety and Health Act of 1970 establishes a set of criteria that the employer

229

will use in protecting his employees against health hazards and harmful materials. Section 6(b)(5) of the Act requires:

The Secretary [of Labor], in promulgating standards dealing with *toxic* materials or *harmful physical agents* under this subsection [of the Act], shall set the standard which most adequately assures to the extent feasible, on the basis of the best available evidence, that no employee will suffer material impairment of health or functional capacity even if such employee has regular exposure to the hazard dealt with by such standard for the period of his working life."

Basically, this establishes guidelines in order to make certain that such standards contain all of the substantive data that relate to the issue or subject covered by the standards so that the employer can better understand compliance requirements. Among such requirements, but not limited thereto, are rules prescribing the use of labels and other appropriate forms of warning as are necessary so that employees can be apprised of all hazards to which they may be exposed, relevant symptoms and appropriate emergency treatment, and proper conditions and precautions of safe use or exposure. Also, where appropriate, such standards must also prescribe suitable protection equipment and control or technological procedures to be used in connection with such hazards and must provide for monitoring or measuring employee exposure at such locations and intervals, and in such manner as may be necessary for protection of employees. Where appropriate, any such

standard shall also prescribe the type and frequency of medical examinations or other tests which shall be made available, by the employer or at his cost, to employees exposed to such hazards in order to most effectively determine whether the health of such employees is adversely affected by such exposures. Also, the OSHAct, Section 8(c)(3), requires the Occupational Safety and Health Administration (OSHA) to (*a*) issue regulations requiring employers to maintain accurate records of employee exposures to potentially toxic materials or harmful physical agents which are required to be monitored or measured, (*b*) provide employees or their representatives with an opportunity to observe such monitoring or measuring and to have access to these records, and (*c*) make appropriate provisions for each employee or former employee to have access to such records as will indicate his own exposure to toxic or harmful physical agents.

In addition, the standard's requirements must stipulate that each employer shall promptly notify any employee who has been or is being exposed to toxic materials or harmful physical agents in concentrations or at levels which exceed those prescribed by any standard, and (the employer) must inform any employee who is being thus exposed of the corrective action being taken.

The employer should be aware of the fact that the OSHA regulations covering occupational health deal with environmental controls. These regulations deal with air contaminants (gases, fumes, vapors, dusts, and mists), noise, and ionizing radiation. It becomes apparent that employers dealing with the aforementioned exposures must, of necessity, become familiar with industrial hygiene concepts and controls.

Industrial Hygiene Concepts

Key concepts

The success of the employer's compliance efforts will be predicated on how diligently the three key concepts for an effective industrial hygiene program are applied. These concepts are:

Recognition—which requires knowledge of stresses arising out of work operations and processes

Evaluation—judgment or decision usually involving measurement of magnitude of stress and comparison with health standards

Control—by isolation, substitution, change of process, wet methods, local exhaust ventilation, general or dilution ventilation, personal protective equipment, housekeeping, and training and education

Physical classification of airborne contaminants

Use the precise term. Both the professional and the nonprofessional must be aware of the precise meanings of certain words commonly used in industrial hygiene if he is going to : (*a*) understand the requirements of the OSHA regulations, (*b*) effectively communicate with other workers in this area, and (*c*) intelligently prepare purchase orders for the procurement of health services

and personal protective equipment. A fume respirator, for instance, is worthless as protection against gases or vapors. Too frequently, terms (such as gases, vapors, fumes, and mists) are used interchangeably. Each term has a definite meaning and describes a certain state of matter which can be achieved only by certain physical changes to the substance itself. Basic to the industrial hygiene vocabulary are states of matter. These are defined in the physical classifications that follow.

Physical classifications.

Dusts. These are solid particles generated by handling, crushing, grinding, rapid impact, detonation, and decrepitation (breaking apart by heating) of organic or inorganic materials, such as rock, ore, metal, coal, wood, and grain.

Dust is a term used in industry to describe airborne solid particles that range in size from 0.1 to $25\,\mu$ ($1\,\mu = 0.0001$ centimeter or $1/25,400$ in.; μ is the abbreviation for micron). Dusts above $5\,\mu$ in size usually will not remain airborne long enough to present an inhalation problem.

Dust may enter the air from various sources. It may disperse when a dusty material is handled, such as when lead oxide is dumped into a mixer or when talc is dusted on a product. When solid materials are reduced to small sizes in processes such as grinding, crushing, blasting, shaking, and drilling the mechanical action of the grinding or shaking device supplies energy to disperse the dust formed.

Fumes. A fume is formed when a volatilized solid, such as a metal, condenses in cool air. The solid particles that make up a fume are extremely fine—usually less than 1.0μ. In most cases, the hot material reacts with the air to form an oxide. Examples are lead oxide fume from smelting and iron oxide fume from arc welding. A fume can also be formed when a material such as magnesium metal is burned or when welding or gas cutting is done on galvanized metal.

Gases and vapors are *not* fumes, even if newspaper reporters often (and incorrectly) call them that.

Smoke. Carbon or soot particles less than $0.1\ \mu$ in size, which result from the incomplete combustion of carbonaceous materials such as coal or oil, are called smokes. Smoke generally contains droplets as well as dry particles. Tobacco, for instance, produces a wet smoke composed of minute tarry droplets.

Aerosols. Liquid droplets or solid particles dispersed in air, that are of fine enough particle size to remain so dispersed for a period of time, are called aerosols.

Mists. Mists are suspended liquid droplets generated by condensation from the gaseous to the liquid state or by breaking up a liquid into a dispersed state, such as by splashing, foaming, or atomizing. Mist is formed when a finely divided liquid is suspended in the atmosphere. Examples are the oil mist produced during cutting and grinding operations, acid mists from electroplating, acid or alkali mists from pickling

operations, paint spray mist from spraying operations, and the condensation of water vapor to form a fog or rain.

Gases. Normally gases are formless fluids which occupy the space or enclosure in which they are confined and which can be changed to the liquid or solid state only by the combined effect of increased pressure and decreased temperature. Examples are arc welding gases, internal combustion engine exhausts gases, and air.

Vapors. The gaseous form of substances that are normally in the solid or liquid state (at room temperature and pressure) are called vapors. The vapor can be changed back to the solid or liquid state either by increasing the pressure or decreasing the temperature alone. Evaporation is the process by which a liquid is changed into the vapor state and mixed with the surrounding atmosphere. Solvents with low boiling points will volatilize readily.

Related hazards. If a compound is very soluble—such as ammonia, formaldehyde, sulfuric acid, or hydrochloric acid—it is rapidly absorbed in the upper respiratory tract and does not penetrate deeply into the lungs. Consequently the nose and throat become so irritated that a person is driven out of the exposure area.

Compounds that are insoluble in body fluids cause considerably less throat irritation than the soluble ones, and may penetrate deeply into the lungs. Thus, a very serious hazard can be present and not be immediately recognized. Examples of such gases are nitrogen dioxide and ozone. The immediate danger from these compounds in high concentrations is acute lung irritation or, possibly later, pneumonia.

There are numerous chemical compounds that do not follow the general solubility rule. Such compounds are not very soluble in water and yet are very irritating to the eyes and respiratory tract. They also can cause lung damage and even death under the right conditions. An example is acrolein.

Both professional and non-professionals recognize that air contaminants exist as a gas, dust, fume, mist or vapor in the workroom air. In evaluating the degree of exposure, the measured concentration of the air contaminant is compared to the published standards on limits of exposure. The American Conference of Governmental Industrial Hygienists adopts a list of *threshold limit values* for about 400 substances on an annual basis.*

The term used to designate the amount by volume of flammable vapors or gases in the atmosphere is *percent.* The lower flammable limit of carbon monoxide gas, for example, is 12.5 percent—a fatal concentration if inhaled. One part of carbon monoxide gas in 99 parts of air, resulting in a 1 percent mixture, would be in the safe range with regard to a fire hazard, but would still be deadly in terms of a health hazard. The point here is that, in most cases, by protecting against the health hazard, the safety professional would elimi-

nate the fire and explosion hazard.

Recognition of Health Hazards

Principal categories of hazards

In order to recognize potential industrial hygiene problems, health hazards (often referred to as environmental stresses) are grouped into the following categories:

Chemical—liquids, gases, dusts, fumes, mists, and vapors as air contaminants and skin irritants.

Physical—radiations, noise, vibration, and extremes of temperatures and pressure.

Biological—insects, molds, fungi, and bacteria.

Ergonomic—work environment, physical and mental stress and fatigue.

Chemical hazards

Mode of entry. Chemical compounds in the form of liquids, gases, mists, dusts, fumes, and vapors may cause problems by inhalation (breathing), by absorption (through direct contact with the skin), or by ingestion (eating or drinking).

Inhalation. Contaminants that can be inhaled into the lungs can be physically classified as gases, vapors, and particulate matter. Particulate matter can be further

---

*OSHA has adopted this listing of Threshold Limit Values in its regulations. It is also published in NSC's *Accident Prevention Manual* and in *Fundamentals of Industrial Hygiene.*

classified as a dust, fume, smoke, aerosol, or mist.

Absorption. Absorption through the skin can occur quite rapidly if the skin is cut or abraded; intact skin, however, offers a reasonably good barrier to chemicals. Unfortunately, there are many compounds that may be absorbed through intact skin. Some are absorbed by way of the hair follicles and others dissolve in the fats and oils of the skin, like organic lead compounds, many nitro compounds, and organic phosphates such as some pesticides.

Compounds that are good solvents for fats (such as toluene and xylene) may cause problems by being absorbed through the skin.

Ingestion. People do not knowingly eat or drink harmful chemicals. Toxic compounds are capable of being absorbed from the gastrointestinal tract into the blood. Lead oxide can cause serious problems if people working with this material are allowed to eat or smoke in work areas. Also, careful and thorough washing is required before eating and at the end of every shift.

Significant considerations. Inhalation as a route of entry is particularly important because of the rapidity with which a toxic material can be absorbed in the lungs and pass into the blood stream. It is important that all routes of entry be studied when an evaluation of the work environment is being made; candy bars, lunches, and smoking in work areas. Personal hygiene and sanitation (in addition to airborne contaminants in work areas) must be considered as additional source of contamination.

Recognition factors. To determine environmental factors or stresses, one must first recognize the chemical makeup of raw materials and the nature of the products and by-products manufactured. This sometimes requires considerable effort. Labels, hazardous material data sheets, and other sources of information are available to assist in this connection. (See NSC's *Fundamentals of Industrial Hygiene.*)

Many industrial materials such as plastics are relatively inert and nontoxic under normal conditions of use, but when heated or machined, they may decompose to form highly toxic by-products. Information concerning products and by-products can be obtained from suppliers' chemical departments.

One can obtain much valuable information by observing the manner in which these health hazards are generated, by the number of people involved, and the control measures in use.

After the list of chemicals and physical conditions to which employees are exposed has been prepared, it is necessary to determine which chemicals or agents result in hazardous exposures and need further study.

Dangerous materials may be defined as substances that may, under specific circumstances, cause injury to persons or damage to property because of reactivity, instability, spontaneous decom-

234

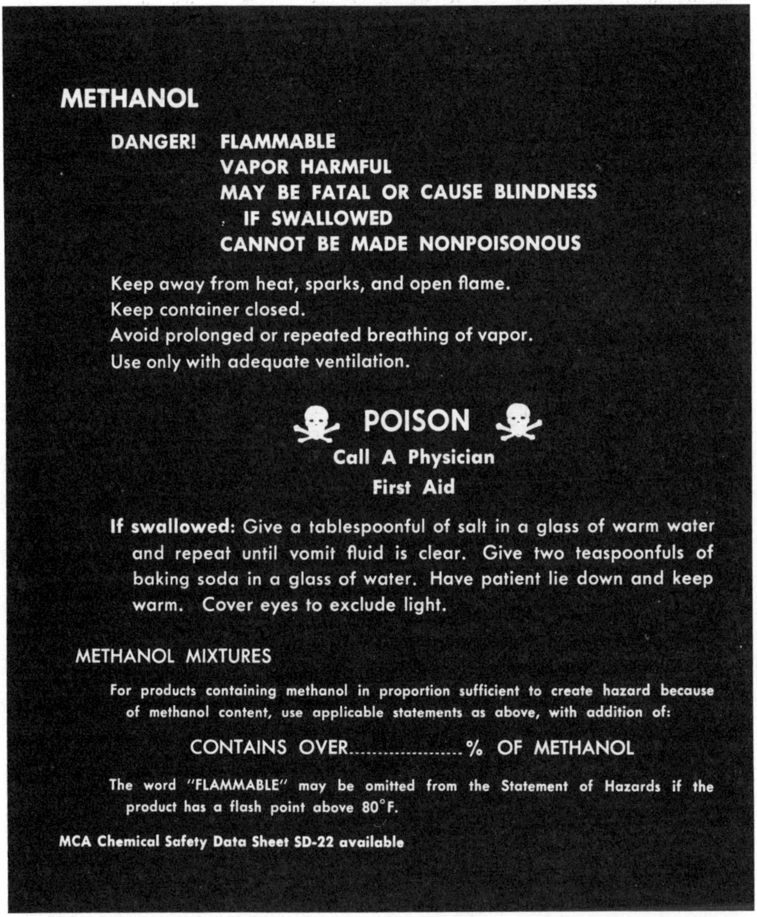

METHANOL

DANGER! FLAMMABLE
VAPOR HARMFUL
MAY BE FATAL OR CAUSE BLINDNESS
IF SWALLOWED
CANNOT BE MADE NONPOISONOUS

Keep away from heat, sparks, and open flame.
Keep container closed.
Avoid prolonged or repeated breathing of vapor.
Use only with adequate ventilation.

☠ POISON ☠
Call A Physician
First Aid

If swallowed: Give a tablespoonful of salt in a glass of warm water and repeat until vomit fluid is clear. Give two teaspoonfuls of baking soda in a glass of water. Have patient lie down and keep warm. Cover eyes to exclude light.

METHANOL MIXTURES

For products containing methanol in proportion sufficient to create hazard because of methanol content, use applicable statements as above, with addition of:

CONTAINS OVER.................% OF METHANOL

The word "FLAMMABLE" may be omitted from the Statement of Hazards if the product has a flash point above 80°F.

MCA Chemical Safety Data Sheet SD-22 available

Courtesy Manufacturing Chemists' Assn.

Figure 13-1.—Typical hazardous chemical label. For labeling of consumer products and identification during transporation, consult appropriate federal agencies.

position, flammability, or volatility.

Explosives are defined as those substances, mixtures, or compounds which are capable of producing an explosion.

Corrosives are those which are capable of destroying living tissue and have a destructive effect on other substances, particularly on combustible materials which may result in a fire or explosion.

Flammable liquids are defined as those liquids with a flash point of 100 F (38 C) or less, although those with higher flash points may be dangerous.

Toxic chemicals are defined as

(Text continues on page 238.)

HAZARDOUS MATERIALS INFORMATION SHEET

(Please complete all applicable sections.)

1. Product Name, Number, Synonym _____ Chemical Formula _____

2. Manufacturer's Name _____

3. Manufacturer's Address _____

4. Chemical and Physical Properties: a. Molecular Wt. _____ b. Boiling Point _____°C

    c. Melting Point _____°C d. Specific Gravity (water = 1) or Bulk Density _____

      @ _____°C e. Vapor Density (air = 1) _____ f. Vapor Pressure (mm Hg) _____

      @_____°C; _____ @ _____°C; _____ @ _____°C;

    g. Solubility _____

    h. pH/conc. _____ i. Index of Refraction _____ @ _____°C

    j. Corrosive action on materials (e.g. aluminum, carbon steel, copper, rubber, plastics, etc.) _____

    _____

    k. Does the material decompose when exposed to air? water? heat? strong oxidizers? possible

      products? _____

    l. Does the material generate heat through polymerization or condensation? _____

    m. Composition *(give chemical names of components; information will be treated as confidential)*

COMPOUND	PERCENT	COMPOUND	PERCENT

NOTE: *Please be specific. For example, it is important to know whether an alcohol is methanol; an aromatic hydrocarbon is benzene; a chlorinated material is carbon tetrachloride.*

5. Flammability and Explosive Properties: a. Flash Point, F, Closed Cup _____

    Open Cup _____ If flash point changes during evaporation give data _____

    b. Explosive limits (% by vol. in air): LOWER _____ UPPER _____

    c. Susceptibility to spontaneous heating: YES _____ NO _____

    d. Fire point, F _____ Auto-ignition temp., F _____

    e. What products might be formed in the event of fire or abnormal temperatures? _____

    f. Suitable extinguishing agents _____

Figure 13-2.—Typical form for

6. Procedures in Case of Container Breakage or Leakage _____

7. Transportation and Storage Requirements _____

8. Physiological Properties *(give animal tested, observation time, dosage value and range, dilution medium, etc.):*

   a. Acute oral toxicity _____

   b. Acute local effects on eyes _____

   c. Acute local effects on skin. Primary irritant? _____

   Sensitizer? _____

   d. Acute inhalation toxicity *(vapor, mist, fume, dust. Indicate effects of concentration and time.)*

   _____

   e. Chronic effects _____

   f. Warning properties *(odor; irritation of eyes, nose, throat)* _____

   g. Threshold limit value *(estimate, if not on current list of ACGIH)* _____

9. First Aid Treatment:

   a. Skin contact _____

   b. Eye contact _____

   c. Inhalation _____

   d. Antidote and treatment in case of swallowing _____

10. Recommended Pre-placement or Periodic Medical Examination *(health standards, clinical tests, frequency, etc.)* _____

11. Precautions for Normal Conditions of Use _____

12. Recommended Personal Protective Equipment _____

13. Suggested Method for Air Analysis _____

14. Pertinent Literature References _____

15. Information Furnished By: NAME _____ DATE _____
    TITLE _____
    COMPANY _____
    ADDRESS _____

(If more space is needed for comment, please attach an additional sheet. Please attach product information data sheets or other publications related to the safe handling and use of this material.)

getting more information from a supplier.

those gases, liquids, or solids which through their chemical properties can produce injurious or lethal effects upon contact with body cells. The majority of toxic chemicals are safe when packaged in their original shipping containers or contained in a closed system.

Oxidizing materials are defined as those chemicals which will decompose readily under certain conditions to yield oxygen. They may cause a fire in contact with combustible materials, may react violently with water, and when involved in a fire may react violently.

Dangerous gases are defined as those gases which may cause lethal or injurious effects and damage to property by their toxic, corrosive, flammable, or explosive physical and chemical properties.

Storage of dangerous chemicals at the location where they are used should be limited to one day's supply, consistent with the safe and efficient operation of the process. The storage should also comply with local ordinances. An approved storehouse should be provided for the main supply of hazardous materials.

It therefore becomes essential to have all possible available data and information concerning dangerous chemicals used or handled. Particularly for purposes of evaluating their harmful effects with regard to employee exposure and their dangerous potential with respect to storage, labels and hazardous material information sheets have been found to be most helpful.

The required information frequently can be obtained from labels on drums and containers (Figure 13-1). However, if the labels do not give complete information but only trade names, it may be necessary to correspond with the manufacturer of the chemicals for the information.

A hazardous materials information sheet, similar to the sample shown in Figure 13-2, can be prepared to elicit toxicological information from the supplier. The information in a completely answered questionnaire can be useful to the medical, purchasing, managerial, engineering, and safety departments in setting guidelines for safe use of these materials in industry. This information would be very helpful in the event of an emergency. Information should be sought on those materials actually in use, together with those which may be comtemplated for early future use. Possibly the best and earliest source of information concerning such materials is to have the manufacturer or distributor provide such data as a purchase contract requirement. Accordingly, it is strongly recommended that a close liaison be set up between purchasing and health and safety so that early information will be provided concerning materials in use and those to be ordered.

Solvents. Although some general hazards and recognition factors arising from the use of solvents are discussed here; a more detailed discussion will be presented in this chapter under Selected Hazards and Methods of

238

Control, pages 260–263.

The widespread industrial use of solvents presents a major problem to the industrial hygienist, the safety professional, and others charged with the responsibility for maintaining a safe, healthful working environment. When working with solvents, getting the job done without hazard to employees or property is dependent upon the proper selection, application, handling, and control of solvents and an understanding of their properties.

The term *solvent* refers to those organic liquids commonly used to dissolve other organic materials. It includes materials such as naphtha, mineral spirits, gasoline, turpentine, benzene, alcohol, and trichloroethylene. A good working knowledge of the physical properties, nomenclature, and effects of over exposure is very helpful in making a proper assesssment of a solvent exposure. Nomenclature can be misleading. For example, "benzine" is sometimes referred to by the worker as "benzene," a completely different solvent. Some commercial grades of benzine may contain benezene as a contaminant. It is a good policy to verify from the label (refer to Figure 13-1), or from the manufacturer, the specific name and composition of the solvent involved. Manufacturers will usually provide information on the composition of their trade name materials if a confidential request is made.

Particulates. Dusts and particulate matter such as fumes also fall in the chemical category. To eval-uate dust exposures requires knowledge of the chemical composition, particle size, dust concentration in air, method of dispersion, and many other factors.

With the exception of some fibrous materials, dust particles must usually be smaller than 5μ in order to enter the alveoli or inner recess of the lungs. Although a few particles up to 10μ in size may enter the lungs, nearly all the larger particles are trapped in the nasal passages, throat, larynx, trachea, and bronchi, from which they are expectorated or swallowed into the digestive tract.

A person with normal eyesight can detect dust particles as small as 50μ in diameter. Dust of respirable size (below 10μ can be seen only with the aid of a microscope. Most industrial dust consist of particles that vary widely in size, with the small particles greatly outnumbering the large ones. Consequently, with few exceptions, when dust can be seen in the air around an operation, probably more invisible dust particles than visible ones are present.

Dispersion. As a rough approximation, macroscopic particles (those large enough to be visible to the naked eye) are considered to be dispersed by dynamic projection. Microscopic particles (those visible only through a microscope) are considered to have a mass so small that their movement is dependent on the containing air mass. Contaminants such as the larger dust particles, mists, and sprays, which are dispersed by dynamic projection, can cause

239

external injuries such as acid burns, eye damage, and dermatitis. The microscopic particles may be dangerous to health if inhaled.

Nonmetallic dusts. A number of occupational diseases result from exposures to nonmetallic dusts, such as silica and asbestos dusts. Although specialized knowledge and instruments are needed to tell how severe a hazard may be, the alert employer or supervisor can, however, recognize a danger spot and ask for expert help. If he doesn't, a lot of harm can be done. Here are some of the things to watch for.

A process that produces dust fine enough to remain suspended in the air long enough to be breathed should be regarded as questionable until it can be proved safe. Silicosis, for example, is produced in hard rock mines and by quarrying and dressing granite, which contains free silica.

Processes in which materials are crushed, ground, or transported are potential sources of dust. They should either be controlled by wet methods or should be enclosed and ventilated by local exhaust. Points where conveyors are loaded or discharged, transfer points along the conveying system, and heads or boots of elevators should be enclosed and, usually exhaust ventilated. Powdered material that is simply riding along on a belt or other conveyor should cause no dust problem.

The supervisor must check that someone does not cancel the effectiveness of built-in dust controls by tampering with them or by using them improperly. He must also check closely that required respiratory equipment is worn by those workers who need supplementary protection as may be required by OSHA regulations (29 CFR 1910.93, Air Contaminants, and 29 CFR 1910.134, Respiratory Protection).

Fumes. Welding, metalizing, and other operations involving molten metals may produce fumes, which may be harmful under certain conditions.

Arc welding volatilizes metal which condenses—as the metal or its oxide—in the air around the arc. In addition, the rod coating is in part volatilized. These fumes, because they are extremely fine, are readily inhaled.

Iron pigment in the lungs appears to have no effect in producing illness or disability. Its presence shows up in X rays of the lungs and may lead to a mistaken diagnosis of silicosis.

More-toxic fumes—such as those formed when welding structures that have been painted with red lead, or when welding galvanized metal—may produce severe symptoms of toxicity rather rapidly, unless fumes are controlled with good, local, exhaust ventilation, or the welder is protected by respiratory protective equipment.

Most soldering operations, fortunately, do not require temperatures high enough to volatilize an appreciable amount of lead. However, the lead in molten solder pots is oxidized by contact with air at the surface. If this

240

oxide, often called dross, is mechanically dispersed into the air, it may produce a severe lead poisoning hazard.

In operations where this might happen, such as soldering or lead battery making, prevention of occupational poisoning is largely a matter of scrupulously clean housekeeping to prevent the lead oxide from even becoming dispersed into the air. It is customary to enclose melting pots, dross boxes, and similar operations, and to ventilate them well in order to control the hazard.

Toxicity vs. hazard. The toxicity of a material is not synonymous with its health hazard. Toxicity is the capacity of a material to produce injury or harm. Hazard is the possibility that a material will cause injury when a specific quantity is used under certain conditions.

The key elements to be considered when evaluating a material health hazard are: (*a*) How much of the material is required to be in contact with a body cell to produce injury? (*b*) What is the probability that the material being absorbed by the body will result in an injury? (*c*) What is the rate of generation of airborne contaminant? (*d*) What is the total time of contact and (*e*) What control measures are in use?

Actual toxicity is dependent on dose, rate, method, and site of absorption, general state of health, individual differences, diet, and temperature.

Threshold limit values. Limits, called threshold limit values (abbreviated as TLV), have been established for airborne concentrations of many chemical compounds. Since the employer or supervisor, or other responsible person could be involved in discussing problems with employee and industrial hygienists who are conducting surveys, one should understand something about TLV's and the terminology in which their concentrations are expressed.

The basic idea of TLV's is fairly simple. They refer to airborne concentrations of substances and represent conditions under which it is believed that nearly all workers may be repeatedly exposed, day after day, without adverse effect. Because individual susceptibility varies widely, exposure of an occasional individual at (or even below) the threshold limit may not present discomfort, aggravation of a preexisting condition, or occupational illness. In addition to the TLV's set for chemical compounds, there are limits for physical agents, such as noise, microwaves, and heat stress.

The TLV may be a time-weighted average figure which would be acceptable for an 8 hour exposure. For some substances, such as an extremely irritating one, a time-weighted average concentration would not be acceptable, so a "Ceiling" value is established. In other words, a Ceiling limit means that at no time during the 8 hour work period should the airborne concentration exceed the limit.

Establishment of values. Threshold Limit Values have

241

been established for about four hundred substances. A group of well qualified toxicologists, hygienists, and doctors reviews the list annually and values are revised as necessary. OSHA, in its regulations for air contaminants, has adopted, in part, "Threshold Limit Values of Airborne Contaminants" of the ACGIH (American Conference of Governmental Industrial Hygienists).

The data for establishing a TLV comes from animal studies, human studies, and industrial experience. The limit may be selected for one of several reasons. It may be based on the fact that a substance is very irritating to the majority of people exposed to concentrations above a given level. The substances may be an asphyxiant. Other reasons for establishing a limit might be that the chemical compound is anesthetic, or fibrogenic, or can cause allergic reactions, or malignancy. Some TLV's are established because above a certain airborne concentration, a nuisance exists.

The concentration of airborne materials capable of causing problems are quite small. Consequently the hygienist uses special terminology to define these concentrations. He often talks in terms of parts per million (ppm) when describing the airborne concentration of a gas or vapor. If he is measuring airborne particulate matter, such as a dust or fume, he uses the term milligrams per cubic meter (mgs/M^3) to define concentrations. He also may use the actual number of particles present in one cubic foot of air

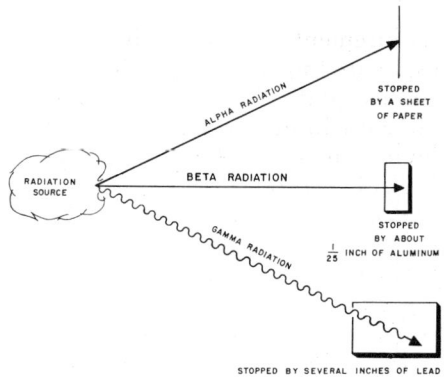

Reprinted with permission from "Atomic Radiation" ©1957 by RCA Service Co., Inc.

Figure 13-3.—The penetrating power of alpha, beta, and gamma radiation.

being tested, in which case the descriptive term become millions of particles per cubic foot (mppcf). As an example of the very small concentrations involved, the hygienist commonly samples and measures substances in the air of the working environment in concentrations ranging from 1 to 100 ppm. Some idea of the magnitude of these concentrations can be appreciated when one realizes that 1 in. in 16 miles is a part per million, one cent in $10,000, one ounce of salt in 62,500 lbs of sugar, one ounce of oil in 7812.5 gal of water—all represent one part per million.

OSHA requirements. Under the OSHA regulations (29 CFR 1910.93, Air Contaminants), an employee's exposure to any material listed in its Threshold Limit Value Tables G-1, G-2, or G-3 must be limited in accordance with the requirements specified therefor. Reference should be made to these tables and their requirements for materials whose exposure limits are set by "ceiling values" and for those other material with limits prescribed for "8 hour time-weighted averages." Care should be exercised in identifying those materials (Table G-2) whose exposure is further restricted by maximum duration (in minutes) above acceptable ceiling concentrations for an 8 hour shift. Table G-3 lists and limits employee exposure to "mineral dusts" an 8 hour work shift of a 40 hour week and not to exceed the 8 hour time-weighted average limit given for the particular material.

Physical agents and related hazards

The next major area of environmental factor or stresses mentioned under principle categories of hazards involves physical agents. Problems relating to such things as noise, ionizing radiation, nonionizing radiation, temperature extremes, and pressure extremes fall in this category. It is important that the employer, supervisor, and/or those responsible for safety and health be alert to the hazards because they may have immediate or cumulative effects on the health of the employees in the workplace.

Noise. Noise (defined as "unwanted sound") is a form of vibration which may be conducted through solids, liquids, or gases. The effects of noise on man include the following:

• Psychological effects (noise can startle, annoy, and disrupt concentration, sleep, or relaxation).

• Interference with communication by speech, and as a consequence, interference with job performance and safety.

• Physiological effects (for example: noise induced loss of hearing, or aural pain, even nausea when the exposure is severe).

Because of the complex relationships of noise and exposure time to partial loss of hearing and the many other possible contributory causes for hearing loss, it is difficult to give exact rules for protecting workers against hearing loss. However, criteria have been developed to protect against hearing loss in the speech frequency range. In fact, these criteria, known as the Threshold Limit Values for noise, are observed by the U.S. Department of Labor under the Occupational Safety and Health Act; they have been promulgated under OSHA regulations 29 CFR 1910.95. More detailed information on the noise requirements is given later in this chapter under "Selected hazards and methods of control."

Ionizing radiation. Although ionizing radiation is a complex subject, a brief discussion is included here because it is one of the more important physical agents.

Gamma radiation from radioactive materials and X radiation are highly penetrating and may produce damage in any tissue of the body. What tissue will be damaged depends in part on the energy of the radiation and in part on the relative sensitivity of the tissue.

243

Alpha and beta radiation are generally also considered with gamma and X radiation, although they are not as penetrating. (See Figure 13-3.) Their danger arises when the materials which produce them have been ingested and fixed in some tissue where the radiations can carry on their destruction at close range.

These types of radiation all cause injury by ionizing the tissue in which they are absorbed. Such injuries differ widely in locations and extent but the determining factors are the ability of the particular radiation to penetrate through the tissues and the amount of ionization produced by a given physical amount of the radiation.

X radiation is produced by high-potential electrical discharge in a vacuum and should be anticipated in evacuated electrical apparatus operating at a potential of 10,000 volts or more.

Employers utilizing ionizing radiation must comply with the OSHA regulations 29 CFR 1910.96, which establishes, among other requiremens, specific limits of exposure of individuals to radiation in restricted areas, precautionary procedures and personal monitoring, use of caution signs, labels, and signals, maintenance of employee exposure records, and reporting of incidents of excessive exposure and concentrations, and damage to property.

Nonionizing radiation. Electromagnetic radiation has varying effects on the body, depending largely on the particular wave-

244

length of the radiation involved. Following, in approximate order of decreasing wavelength and increasing frequency, are some hazards associated with different regions of the nonionizing electromagnetic radiation spectrum. Nonionizing radiation is covered in detail by OSHA regulations under 29 CFR 1910.97, and in NSC's Fundamentals of Industrial Hygiene.

Low frequency. The longer wavelengths—including power frequencies, broadcast radio, short wave radio—can produce general heating of the body. The health hazard from these radiations is very small, however, since it is unlikely that they would be found in intensities great enough to cause significant effect.

Microwaves have wavelengths of 3 m to 3 mm (100 to 100,000 megahertz, MHz). They are found in radar, communications, and diathermy applications. Microwave intensities may be sufficient to cause significant heating of tissues.

An intolerable rise in body temperature, as well as localized damage, can result from an exposure of sufficient intensity and time. In addition, flammable gases and vapors may ignite when they are inside metallic objects located in a microwave beam.

Infrared radiation does not penetrate below the superficial layer of the skin so that its only effect is to heat the skin and the tissues immediately below it. Except for thermal burns, the health hazard upon exposure to infrared radiation sources is negligible.

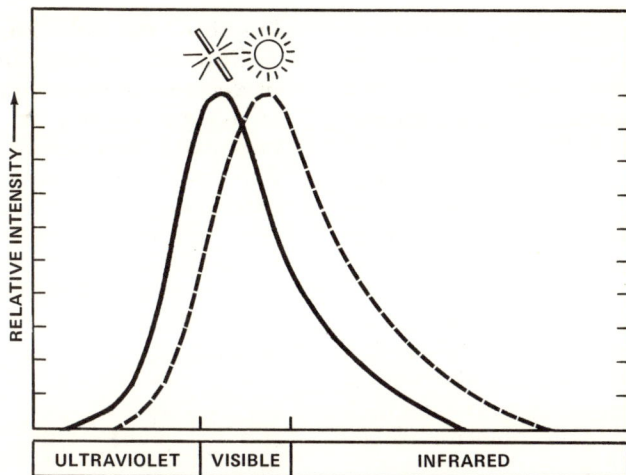

Figure 13-4.—Energy distribution in the radiation from a welding arc (solid line) and the sun (dashed line).

Visible radiation and ultraviolet radiation also does not penetrate appreciably below the skin; their effects are essentially heating on the surface. The effects of the ultraviolet (UV) waves are much more violent than are those of the visible and infrared—a severe burn can be produced with no warning whatever and significant damage to the lens of the eye can occur from excessive exposure. (See Figure 13-4.)

Since ultraviolet radiation in industry may be found around electrical arcs, they should be shielded by materials opaque to the ultraviolet. Opacity to ultraviolet has no relation to opacity to visible part of the spectrum. Ordinary window glass, for instance, is almost completely opaque to the ultraviolet although transparent to the visible. A piece of plastic dyed a deep red-violet may be almost entirely opaque in the visible part of the spectrum and transparent in the near ultraviolet.

Electric welding arcs and germicidal lamps are the most common strong producers of ultraviolet in industry. The ordinary fluorescent lamp generates a good deal of ultraviolet inside the bulb, but it is essentially absorbed by the bulb and its coating.

The most common exposure to ultraviolet radiation is from direct sunlight, and a familiar result of overexposure—known to all bathers—is sunburn. Most everyone is also familiar with certain compounds and lotions which minimize the effects of the sun's rays, but many are unaware that some industrial materials, such as cresols, make the skin expecially sensitive to ultraviolet rays. So much so that after having been exposed to cresols, even a short exposure in the sun usually results in a severe sunburn.

245

Visible radiations. Visible radiation, which is about halfway up the spectrum of frequency and wavelength, is important because it can effect both the quality and accuracy of work. Good lighting invariably results in increased product quality with less spoilage and increased production. Good lighting contributes to sanitary, clean, and neat operations.

Lighting should be bright enough to produce easy seeing and directed so that it does not create glare. The brightness level should be high enough to permit efficient seeing. Illumination levels and brightness ratios recommended for manufacturing and service industries are recommended by the Illuminating Engineering Society.* OSHA has not, to date, adopted the ANSI A11.1 Standard per se. It has however, adopted many other ANSI Standards in its General Industry Standards (29 CFR 1910) that adopt ANSI A11.1 by reference. Such adoptions have the force and effect of law, where such standards apply.

One of the most objectionable features of lighting is glare—brightness within the field of vision that causes discomfort or interferes with seeing. The brightness can be caused by either direct or reflected light. To prevent glare, keep the source of light well above the line of vision, or shield it with opaque or translucent material.

Almost as bad, is an area of excessively high brightness in the visual field. A highly reflecting white paper in the center of a dark, nonreflecting surface, or a brightly illuminated control handle on a dark or dirty machine are two examples.

To prevent this condition, keep surfaces light or dark with little difference in surface reflectivity. Color contrasts are acceptable, however.

Although it is generally best to provide even, shadow-free light, some jobs require contrast lighting. In these cases, keep the general (or background) light well diffused and glareless and add a supplementary source of light that will cast shadows as wanted. Avoid too much contrast in brightness between the work and the surrounding area, and avoid direct or reflected glare.

Lasers. Lasers emit beams of coherent light of a single color (or wavelength). Contrast this with conventional light sources that produce random, disordered lightwave mixtures of various wavelengths. The word *laser* is an acronym for Light Amplification by Stimulated Emission of Radiation. A laser beam is made up of lightwaves that are nearly parallel to each other (collimated) and are all traveling in the same direction.

The maser, the laser's predecessor, emits microwaves instead of light. Some companies call their lasers "optical masers." Uses for the laser include machining and cutting of metals, welding of microscopic parts and systems for

---

*Illuminating Engineering Society, *Practices for Industrial Lighting,* is designated as American Standard A11.1.

high-capacity communications.

Since the laser is highly collimated (has a small divergence angle), it can have a large energy density in a narrow beam. Direct viewing of the laser should therefore be avoided. The work area should contain no reflective surfaces (such as mirrors or highly polished furniture) for even a reflected laser beam can be hazardous. Suitable shielding to contain the laser beam should be provided.

The fact that infrared radiation of certain lasers may not be visible to the naked eye contributes to the potential hazard.* OSHA covers protection against laser hazards in its construction regulations.

Lasers generating in the ultraviolet range of the electromagnetic spectrum can produce corneal burns rather than retinal damage, because of the way the eye handles ultraviolet light.

Other factors which have a bearing on the degree of eye injury induced by laser light are: (*a*) the pupil size—the smaller the pupil diameter, the less the amount of laser energy permitted to the retina, (*b*) the power of the cornea and lens to focus the incident light on the retina, (*c*) the distance from the source of energy to the retina, (*d*) the energy and wavelength of the laser, (*e*)the pigmentation of the eye, (*f*) the place on the retina where the light is focused, (*g*) the divergence of the laser light, and (*h*) the presence of scattering media in the light path.

Temperature extremes—heat.

General experience shows that extremes of temperature affect the work that a person does and the manner in which he does it. The industrial problem is more often that of exposure to high temperatures rather than low temperatures.

The body continuously produces heat. Since the body processes are so designed that they can operate only with a very narrow temperature range, they must dissipate this heat as rapidly as it is produced if the body is to function efficiently. A sensitive and rapidly acting set of thermostatic devices in the body must also control the rates of its temperature-regulating processes.

Radiant heat is electromagnetic energy that does no heating until it strikes some object, like a person, where it is absorbed. No relief from it whatsoever results from blowing air around. Protection can be done by placing any kind of opaque shield or screen between the man and the radiating surface, such as discussed next under Heat controls.

Conduction becomes an important means of heat loss when the body is in contact with a good cooling agent, such as water. For this reason, when people are immersed in cold water, they become chilled much more rapidly and effectively than when ex-

*Chapter 11 of this handbook covers requirements for personal protective equipment for protection against radiation hazards, including lasers.

247

posed to air of the same temperature.

Air movement cools the body by convection, the moving air removes from about the body the saturated air (which is formed very rapidly by evaporation of sweat) and replaces it with a fresh layer, capable of accepting more moisture.

Effects of high temperature. The body attempts to counteract the effects of high temperature by increasing the rate of heart beat. The capillaries in the skin then dilate to bring more blood to the surface so that both the rate of cooling and, gradually, the body temperature are increased. Intermittent rest periods for persons necessarily exposed to extreme heat reduces this danger.

HEATSTROKE (also known as sunstroke) is not necessarily the result of exposure to the sun. It is caused by exposure to an environment in which the body is unable to cool itself sufficiently, with the result that the body temperature rises rapidly.

Heatstroke is a much more serious condition than heat cramps or heat exhaustion. An important predisposing factor is excessive physical exertion. The only method of control is to reduce the temperature of the surroundings or to increase the ability of the body to cool itself, so that body temperature does not rise. See following section on Heat Controls, for discussion on heat shields.

HEAT CRAMPS may result from exposure to high temperature for a relatively long time, particularly if accompanied by heavy exertion, with excessive loss of salt and moisture from the body. Even if the moisture is replaced by drinking plenty of water, an excessive loss of salt may provoke heat cramps or heat exhaustion.

HEAT EXHAUSTION may also result from physical exertion in a hot environment. Its signs are a relatively low temperature, pallor, weak pulse, dizziness, profuse sweating, and cool moist skin.

Experience has shown that a worker does not tolerate a hot job very well at first, but develops a tolerance rapidly and acquires full acclimatization within a week to a month.

Heat controls. A great deal can be done in "hot" industries to improve the health, efficiency, and comfort of workers exposed to high temperature and humidity. Both equipment and good planning are required.

Heat shields and ventilation are two items that can be used. For instance, a large radiant area may heat a person excessively, even if the air surrounding the person is not very hot. In which case, it is more effective to provide a reflector shield to return the radiant energy toward its source than to try to cool the person by an air blast or to try to insulate the radiant source.

Sheet aluminum reflectors completely surrounding (but not touching) a furnace will reflect 95 percent (or more) of the radiant energy, and will themselves absorb so little heat that they will be cooled by natural circulation of the air around them. Not only will these reflectors protect workmen from excessive radiant heat, they will also reduce the atmospheric temperature because they decrease the amount of heat that is received and re-radiated by other objects.

Makeshift remedies like an iron panel, for example, can become heated, and then act as a radiator itself. Fortunately, aluminum foil, sheets, and corrugated siding are all highly reflective in the infrared—and they are inexpensive. See Chapter 11, Personal Protective Equipment, for other special clothing used for protection against radiant heat and extreme temperatures.

Temperature extremes—cold. As evidenced by the preceding discussion, a person must work, at times, in places that are extremely cold or extremely hot. The problems related to each type of exposure has been the subject of many investigations and studies. Although more work has been done on the evaluation of the stress from hot environments than that from cold. this section will serve to highlight some of the effects and controls of cold extremes. Additional information may be obtained from other NSC texts.

Air movement is of somewhat more importance in cold environment than in hot because the combined effect of the air movement and temperature can produce a condition called "wind chill." For example, with a temperature of 10 F above zero and a wind of 20 mph, exposed flesh probably would freeze. Thus, unless clothing protects all body areas against considerably lower temperatures than 10 F, a person working outside could soon be in serious trouble.

When exposed to such conditions, wear good protective clothing such as freezer jackets, insulated underwear under denim work clothes, synthetic fabric pants and jacket combinations worn over or under other clothing, hoods, parkas, and other arctic clothing, and insulated gloves. Make sure clothing does not restrict circulation.

Biological stresses

The next significant category of environmental factors or stresses to be discussed under the principle categories of hazards is of a biological nature and is related to problems with animal handling, insects, molds, fungi, bacteria, and grain dusts.

Fungus infections. A number of occupational infections are common to workers in agriculture and in closely related industrial jobs. Grain handlers who inhale grain dust are likely to come in contact with some of the fungi which contaminate grain from time to time. Some of these fungi can, and do flourish in human lungs, causing a condition called farmer's lung. Fine, easily spread fungi, such as rust and smut, can be contacted once the covering of infested grain is broken and may produce sensitization. Infections from some of these fungi are endemic (of local nature) in the Southwest and are not necessarily industrial.

Byssinosis. Byssinosis occurs

249

in individuals who have experienced prolonged exposure to heavy air concentrations of cotton dust. Flax dust has also been incriminated. The exact mode of action of the cotton dust is unknown, but one or more of these factors may be important: (*a*) toxic action of micro-organisms adherent to the inhaled fibers, (*b*) mechanical irritation from the fibers, and (*c*) allergic stimulation by the inhaled cotton fibers or adherent materials. It takes several years of exposure before manifestations are noticed.

Ergonomics

The last of the four major categories of environmental factors or stresses involves human reaction to monotony, fatigue, repeated motion, and repeated shock.

The term "ergonomics" literally means the customs, habits, and laws of work. According to the International Labor Office it is: " . . . The application of human biological science in conjunction with the engineering sciences to achieve the optimum mutual adjustment of man and his work, the benefits being measured in terms of human efficiency and well-being."

In the broad sense, the benefits which can be expected from designing work systems to minimize physical and mental stress on workers are:

- More efficient operation
- Fewer accidents
- Lower cost of operation
- Reduced training time
- More effective use of personnel

The human body can endure considerable discomfort and stress and can perform many awkward and unnatural movements—for a limited period of time.

However, when abnormal conditions or motions are continued for prolonged periods, the physiological limitations of the worker may be exceeded. To assure a continued high level of performance, work systems must be tailored to human capacities and limitations.

Mechanical vibration. A condition known to stonecutters as "dead fingers" or "white fingers," occurs mainly in the fingers of the hand used to guide a cutting tool. The circulation in this hand becomes impaired. When exposed to cold, the fingers become white and lose sensation as though mildly frostbitten. The condition usually disappears after the fingers have been warmed for a short time. However, some workers experience such a severe disability (although temporary) that they are forced to seek other types of work. Prevention of this condition is much more satisfactory than treatment. Preventive measures include directing the exhaust air from air-driven tools away from the hands so that they will not become unduly chilled, use of handles of a comfortable size for the fingers and, in some instances, substitution of mechanical cleaning methods for hand methods. In many instances, simply preventing the fingers from becoming chilled while at work will eliminate the condition.

Repeated motion. Excessive repetitive motion or repeated shock as in some sorting and assembling jobs may often cause irritation and inflammation of the tendon sheath of the hands and arms. The condition, generally known as tenosynovitis, is painful and disabling. The usual treatment is rest and diathermy. The condition results from stress repeated to excess. It may happen to an employee who has been working at the same job for years if he is suddenly asked to put in considerable overtime. Is most likely to occur to a new employee or to an employee transferred to a new job.

The employees should be closely watched so that if they show signs of soreness in the backs of the hands, wrists, forearms, or shoulders they can be transferred temporarily to other work or the job can be changed to reduce the stress.

Other occupational health exposures

Industrial dermatitis. Skin problems in industry are important enough to be placed in a category by themselves. The individual problem usually can be traced back to exposure to some kind of a chemical compound or to some form of physical abrasion or irritation of the skin. Occupational skin diseases account for about 60 percent of all compensation claims for occupational diseases. Although rarely a direct cause of death, skin disorders cause much discomfort and are often hard to cure.

General causes. Causes of occupational skin diseases are classified as follows:

- Mechanical agents—friction, pressure, trauma.

- Physical agents—heat, cold, radiation.

- Chemical agents—organic and inorganic. (these are subdivided according to their action on the skin as primary irritants or sensitizers.)

- Plant poisons—several hundred plants and woods can cause dermatitis. (The best known example is poison ivy.)

- Biological agents—bacteria, fungi, parasites.

Even substances that are normally harmless will cause irritations of varying severity in some people.

There are two general types of skin reaction—primary irritation dermatitis and sensitization dermatitis. Practically all persons will suffer primary irritation dermatitis from contact with mechanical, physical, or chemical agents. Brief contact with a high concentration of a primary irritant or prolonged exposure to a low concentration will result in inflammation. Allergy is not a factor under these conditions. Sensitization dermatitis, the second type of reaction, is the result of an allergic reaction to a given substance. Once sensitization develops, even small amounts of the material may cause symptoms.

Occupational acne. Cutting fluids are frequently responsible for occupational acne. The process begins with the skin becoming irritated from contact with the

251

oil. Next a blackhead forms. If it becomes infected, oil pimples or boils result that resemble adolescent acne. Dirty oil, worker carelessness, and use of germicides in the fluid increase the possibility of occupational acne.

Basic control measures. Any investigation of occupational skin disease should consider the following:

• Degree of perspiration

• Personal cleanliness

• Pre-existence of skin disorders and allergic states

• Diet

Oxygen deficiency. Reduced atmospheric pressure is not the only condition under which oxygen starvation may occur. Deficiency of oxygen in the atmosphere of confined spaces is commonly experienced in industry. The oxygen content of any tank or other confined space should be checked before entry is made. Instruments such as the oxygen analyzer are commercially available for this purpose. (Measuring instruments are discussed on pages 255 and 256.)

Normal air contains approximately 21 percent oxygen by volume. The first physiologic signs of a deficiency of oxygen (anoxia) are increased rate and depth of breathing. Oxygen concentrations of less than 16 percent by volume cause dizziness, rapid heart beat, and headache. A person should never enter or remain in areas where tests have indicated such low concentrations unless a supplied-air or self-contained res-

piratory equipment is worn. See Chapter 11, the section entitled Respiratory Protection.

Entering tanks and confined spaces. Ordinary jobs involving maintenance and repair of systems for storing and transporting fluids or entering tanks or tunnels for cleaning and repairs are controlled almost entirely by the immediate supervisor, and accordingly he should be particularly knowledgeable of all rules and precautions to assure the safety of individuals required to work in such atmospheres. These safeguards should be meticulously adhered to.

As an example, there should be a standard operating procedure for entering tanks. Such procedures should be consistent with OSHA regulations and augmented by in-house procedures which may enhance the basic OSHA rules. Even if a tank is empty, it may have been closed for some time and developed an oxygen dificiency. It may be unsafe to enter without supplied-air respiratory protection.

It is considered good practice to provide hose masks and blowers for men entering tanks. It is the supervisor's job to make certain that no workman enters a tank without a hose mask and attached lifeline, nor without a man outside to assist him if he encounters serious difficulties. Reference should be made to applicable federal and state regulations. See pages 200-203.

Evaluation of Hazards

Evaluation defined. Evaluation

may be defined as the decision-making process resulting in an opinion as to the degree of health hazard arising out of an industrial operation. Evaluation involves judgment of the magnitude of environmental factors based upon past experience, and by qualitative or quantitative measurement of the chemical, physical, biological, or ergonomic stresses. Evaluation, in the broad sense, also refers to determining the levels of physical and chemical agents arising out of a process for purposes of studying that process and to determining the effectiveness of a given piece of equipment used to control the process.

Job-site considerations. If a plant is going to handle a toxic material, it is necessary to consider all the kinds of unexpected events that can occur, and what precautions need to be taken to prevent atmospheric release of a toxic material. Once these considerations have been studied and cared for during the design of the plant, it is necessary to instruct operating and maintenance personnel on the health hazards and the safety features provided. Only in this way can they be made aware of the possible hazards and the reasons for certain built-in safety features.

The operating and maintenance people should set up a routine procedure (at frequent stated intervals) for testing the industrial hygiene and safety provisions that may normally not be used in ordinary operation.

Basic data and information needs. As pointed out in previous sections of this text, there is no substitute for knowledge concerning the effects of exposure to the materials and by-products arising out of particular operations being performed in a plant or company. Detailed information should be obtained regarding the hazardous materials that are used within a plant, the type of job operation, how the workers are exposed, work pattern, levels of air contaminant, duration of exposure, what control measures are used, and other pertinent information. The earlier section which discussed recognition factors and recognition aids, use of labels, and hazardous materials information sheets will be extremely helpful in gathering essential data and information. The hazard potential of the material is determined not only by its toxicity, but also by the conditions of use—who uses what, where, and how long? Essential to an industrial health hazard evaluation is some knowledge of toxicology and an understanding of the concepts involved in the use of the Threshold Limit Values and the maximum acceptable concentrations which were discussed earlier in this chapter.

Determining exposure levels. To determine exposure levels to atmospheric contaminants, as well as to measure physical energy stresses, the industrial hygienist uses either direct-reading instruments or laboratory analytical methods.

A particular exposure can be evaluated by chemical analyses of air samples, and a step-by-step analysis of the operations can be made to find the areas where the exposure took place. The operational analysis should also determine how the material becomes airborne and is dispersed.

The type and extent of controls will depend upon the evaluation made of the exposure, and the operation that disperses the contaminant.

Air sampling

The importance of the sampling location, the proper time to sample, and the number of samples to be taken during the course of an investigation of the work environment cannot be over stressed. The duration of exposure (as well as intensity or concentration) is required in order to calculate the dosage a man receives.

An adequate number of tests should be taken to define the time-weighted average exposure to be able to relate this to the threshold limit value. Samples must be taken to characterize the peak emissions during various portions of the entire process cycle, in addition to those taken to determine the average level. Air samples collected within an extremely short sampling period may miss peak concentrations which occur between tests.

Other factors to be considered are changes which occur in plant operations during the day or night shifts, contributions to exposure from adjacent operations, and seasonal variations. Another factor to consider—whether the presence of the sampler affects the test result.

Inexperienced personnel, when confronted with an isolated air sample of high concentration, conceivably might shut down a relatively safe operation, resulting in considerable embarrassment and financial loss. In other situations, if the limitations of the sampling method and instruments are not understood, a serious exposure may be overlooked—it is quite possible to obtain low, seemingly safe readings but have an extremely hazardous condition.

Professional judgment is necessary to reach a decision. Knowledge of the operation or process, particularly its relative magnitude at a given time, is essential for good evaluation of exposure. For example, air samples taken during a production cutback would yield low results, not representative of conditions of exposure under normal operation.

Concentrations of atmospheric contaminants fluctuate as a process cycles, as leaks occur, or as misoperation occurs. Do not assume that all operations produce air contaminants at the same steady level all day long.

Prior to making any test of the workroom air, the user must have a thorough knowledge of the nature of the contaminant in question and of the other substances likely to be present in the atmosphere, in order to make a proper choice of sampling method. NSC's *Fundamantals of Industrial Hygiene* presents principle sample collection methods for air contaminants.

In addition to analysis of the air in a workroom atmosphere, other

254

tests such as biochemical tests on body tissues or excreta can give an accurate measure of a worker's exposure. Such tests should be carried out under medical supervision and be interpreted by a competent physician.

What must be measured and evaluated in industrial hygiene sampling is usually the dosage the worker receives—the amount of toxic material he takes into his body. For evaluating employee exposure, measurements should be made on the air in the worker's breathing zone.

The results obtained must be compared with a threshold limit value with a full knowledge of its limitation and consideration of other factors in the situation. Urine assays add to the facts available for making an evaluation of exposure. These, in turn, must be combined with knowledge obtained by the physician directing an adequate medical program. All this serves to emphasize the basic teamwork approach of the industrial hygiene and health program.

Electrical direct-reading instruments comprise those which give a direct reading on a dial. This is a meter that involves electronic circuitry which requires careful maintenance to function properly.

In one type of electrical direct-reading instrument, the air stream is passed over a heated wire. The contaminant in the air stream is oxidized and the resultant rise in temperature (which varies with the concentration of the contaminant) is measured by means of an electrical circuit.

Examples of electrical direct-reading instruments which measure the amount of combustible gas or vapor contaminant are combustible gas indicators, vapor testers, and explosimeters.

There are many such devices varying widely in sensitivity, reliability, and ease of maintenance. Chapter X, Table 10-B, NSC's *Fundamentals of Industrial Hygiene* lists the more common ones used for both laboratory and field work. All require calibration before use. The accuracy of the electrical direct-reading instruments may be affected by improper maintenance, lack of calibration, changes in airflow rates and volumes, and weak batteries. A routine schedule of maintenance and calibration is therefore necessary.

Color-change and stain-length instruments are those which produce a color change in a sensitive chemical through which the air to be tested is drawn. This is a more numerous type of instrument and one which has had a great deal of development in recent years. These devices depend on a comparison of the colors produced against a set of standard colors or the measurement of the length of the stain produced in a tube packed with reacting chemical (Figure 13-5). There are now available for use on many common chemicals: hydrogen sulfide, aromatic hydrocarbons, hydrogen cyanide, carbon monoxide, trichloroethylene, toluene-diisocyanate, mercury vapor, and others. Manufacturers of these tubes often supply lengthy lists of

255

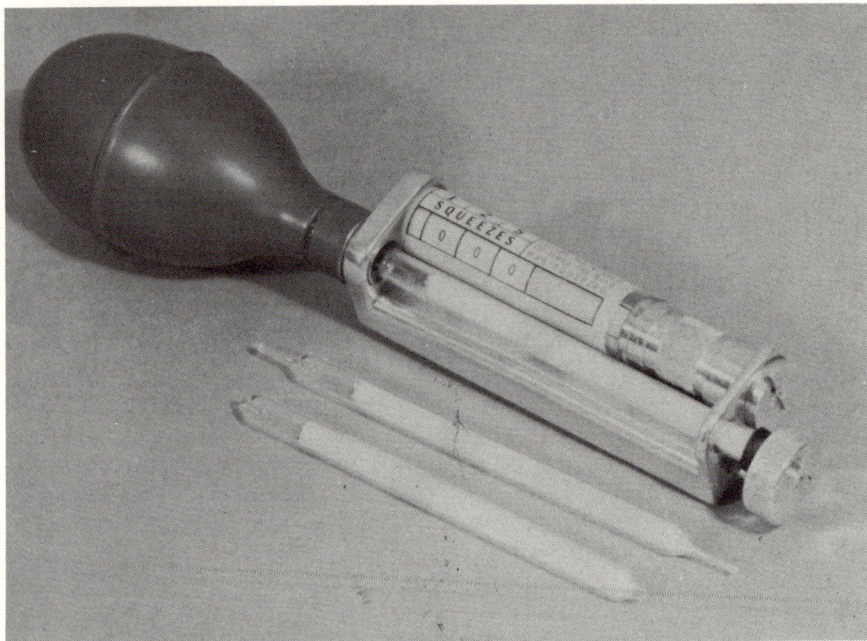

Figure 13-5.—The detecting or collecting medium in the glass indicator tube contains chemically treated granules that will react with the contaminant in the air stream to form a stain or color change. The lengh of the stain, or the intensity of the color (or the first appearance of the color) is a measure of the amount of contaminant present in the air sample.

chemicals which may be detected by such tubes, but the lists sometimes give an exaggerated impression of the versatility of this color-change technique since compounds may be included for which the tubes have only a limited sensitivity.

The colorimetric tubes and papers are deceptively simple to operate. Air is usually drawn through the tube with a hand pump—this means that there is no motor to fail or battery to replace.

Variations, such as nonuniformity and particle size of the media, loss of chemical stability due to long storage, lack of expiration date markings, etc., should be checked. It is the responsibility of the manufacturer to point out all the possible interference with a test, at least to the best of his knowledge. However, it is the responsibility of the user to determine the presence or absence of any interfering airborne substances or other variables which may affect the validity of his measurements in the workroom environment.

Controlling Environmental Hazards

General concepts

The general methods of controlling environmental factors or stresses include the use of engineering controls, administrative controls, and personal protective equipment. Within each of these control methods there are alternatives which can be applied individually or in conjunction with another to achieve the desired level of protection.

Engineering controls

Substitution. Replacement of a toxic material with a harmless one is a very practical method of eliminating an industrial health hazard.

In many cases a solvent with a lower order of toxicity or flammability may be substituted for a more hazardous one. In a solvent substitution, it is always advisable to experiment on a small scale before making the new solvent part of the operation or process. For example, carbon tetrachloride can be replaced by solvents such as methyl chloroform, dichloromethane, aliphatic petroleum hydrocarbons, or one of the fluorochlorohydrocarbons. (The precautions listed in this chapter should be reviewed.) Benzene can be replaced by toluene in most lacquers, synthetic-rubber solutions, and paint removers. Natural rubber cements with aliphatic hydrocarbon solvents can perform virtually the same function as benzene cements.

Changing the process. A change in process often offers an ideal chance to improve working conditions as well as quality and production. In some cases, a process can be modified to reduce the exposure to a dust or fume and thus markedly reduce the hazard. Brush painting or dipping instead of spray painting will minimize the concentration of airborne contaminants from toxic pigments, arc welding in place of riveting, vapor degreasing to replace handwashing of parts, airless spraying techniques, and electrostatic devices for hand spraying. Also in buying individual machines, the need for accessory ventilation, noise suppression, vibration and heat control should be considered before purchase.

Isolation or enclosure. Some potentially dangerous operations can be isolated from the people nearby. The isolation can be by a physical barrier (such as an acoustic box to contain noise from a whining blower or a screaming rip saw).

Isolation is particularly useful for limited operations requiring relatively few workers or where control by any other method is too difficult or expensive.

Enclosing the process or equipment is a desirable method of control since the enclosure will prevent or minimize the escape of solvent vapor into the workroom atmosphere. Where some of the more highly toxic solvents are used, enclosure should be one of the first measures attempted, after considering substitution. Further examples of where this type of control is effective are radium dial painting, glove booths, airless blast or shot blast machines for cleaning castings, metal spraying operations, and abrasive blasting

257

cabinets. In the chemical industry, the isolation of hazardous processes in closed systems is a widespread practice, which is one reason why the manufacture of toxic substances is often less hazardous than their use.

Wet methods. Dust hazards can frequently be minimized or greatly reduced by application of water or other suitable liquid at the source of dust, a method which is often used for silica and loose dusts. Wetting of floors before sweeping to keep down the dispersion of harmful dust is advisable when better methods, such as vacuum cleaning cannot be applied. "Wetting down" is one of the simplest methods for dust control. Its effectiveness, however, depends upon proper wetting of the dust. This may require the addition of a wetting agent to the water and proper dispoal of the wetted dust before it dries out and is redispersed.

Ventilation. Ventilation requirements for welding, burning and cutting were discussed in Chapter 9 of this Handbook. Ventilation requirements for working in confined spaces, oxygen dificient atmospheres, and for working with toxic and hazardous material have been discussed and referenced in earlier sections of this chapter. OSHA ventilation requirements for abrasive blasting; grinding, polishing and buffing operations; spray finishing operations (and use of related solvents); and open surface tanks are covered extensively in 29 CFR 1910.94.

258

Local exhaust ventilation. A local exhaust system traps the air contaminant near its source so that a worker standing at the process is not exposed to harmful concentrations. This method is usually preferred to general ventilation, but should be used only when the contaminant cannot be controlled by substitution, changing the process, isolation, or enclosure. Even though a process has been isolated, it may still require a local exhaust system. More details of this method of control are given in the prior-referenced regulations and in the NSC texts listed in the Foreword.

General or dilution ventilation. General or dilution ventilation—adding air to keep the concentration of a contaminant below hazardous levels—uses natural convection through open doors or window, roof ventilators, and chimneys, or artificial air currents produced by fans or blowers. Exhaust fans through roofs, walls, or windows constitute all season dilution ventilation. Consideration must be given to providing make-up air, especially during winter months. Dilution ventilation is practicable only if the degree of air contamination is not excessive and particularly if the contaminant is released at a substantial distance from the worker's breathing zone. Under other conditions, the contaminated air will not be diluted sufficiently before inhalation. General ventilation should not be used where there are major, localized sources of contamination (especially highly toxic dusts and fumes);

local exhaust is more effective in such cases.

Personal protective equipment

When it is not feasible to render the environment completely safe, it may be necessary to protect the worker from the environment. Personal protective equipment is normally considered to be secondary to the controls mentioned previously and to administrative controls mentioned next. Where it is not possible to enclose or isolate the process or equipment, provide ventilation, or other control measures; where there are short exposures to hazardous concentrations of contaminants; where unavoidable spills may occur—personal protective equipment must be provided and used.

Personal protective devices have one serious drawback—they do nothing to reduce or eliminate the hazard. All such equipment, however, is intended for emergency or temporary use only. Only a brief reference to personal protective equipment is given in this section, because Chapter 11, "Personal Protective Equipment," discusses each type.

Administrative controls

Worker rotation and/or limiting exposures. When the employee's exposure cannot be reduced to permissible safe levels through engineering controls, as in the case of air contaminants or noise, then an effort should be made to limit the employee's exposure through administrative controls. Examples of some administrative controls are:

• Arrangement of work sched-

ules so that employees are minimally exposed to health hazards.

• Assurance that employees who have reached their upper permissible limits of exposure work the remainder of the day in an environment where no further harmful exposures will be experienced.

• Where hours of exposure for a job exceed the permissible exposure time for one man in one day, the work is scheduled for two, three or as many men as are needed to keep each individual's exposure within permissible time limits. In the case of noise, other possibilities may involve part time use of noisy equipment or rescheduling noise producing operations to other shifts where less employees may be exposed.

While not as satisfactory as controlling the offending exposure at its source, administrative control measures are more easily enforced than is the requirement to wear personal protective equipment.

Housekeeping. Good housekeeping plays a key role in occupational health protection. Basically, it is another tool in addition to those already mentioned for preventing dispersion of dangerous contaminants and for maintaining a safe and healthful working condition required by law. Immediate cleanup of any spills of toxic material is a very important control measure. Good housekeeping is also essential where solvents are stored, handled and used. Leaking containers or spigots should be remedied immediately and spills should be cleaned up promptly by workers wearing protection equipment. All solvent-soaked rags or absorbents should be disposed of in airtight metal receptacles and removed daily from the plant.

It is impossible to have an effective health program unless maintenance and housekeeping are good and the worker has been informed for the need for those measures.

Training and education. Proper training and education is an essential element to the successful implementation of any control effort. Workers can be alerted to safe procedures through booklets, instruction signs, labels, safety meetings, and other education devices. (See Chapter 7.)

Selected Hazards and Methods of Control

Noise

The criteria for hearing conservation observed by OSHA in its regulations 29 CFR 1910.95, establishes by law the permissible levels of harmful noise an employee may be subjected to. The permissible exposures at given decibel levels (dBA) and hours (duration per day). For example, a noise of 90 dBA can be tolerated for 8 hours; 95 dBA, for 4 hours.

Basically the regulations stipulate that when employees are subjected to sound exceeding the limits, feasible administrative or engineering controls shall be utilized. If such controls fail to reduce sound exposure within permissible levels, personal protective equipment must be provided and used to reduce sound levels to within the permissible levels. Exposure to impulsive or impact noise should not exceed 140 dB peak sound pressure level.

If the variations in noise level involve maxima at intervals of 1 second or less, it is to be considered continuous.

In all cases where the sound levels exceed the permissible values, a continuing, effective hearing conservation program shall be administered. (NSC's book *Industrial Noise and Hearing Conservation* gives details.)

Determining noise levels. Limits have been set for exposure to noise, but normally it will not be possible for nonskilled persons to measure noise levels. This should be done by an industrial hygienist or a safety professional who then gives those responsible the necessary data for establishing appropriate controls, or advises as to the necessity for hearing protection.

Rule-of-thumb determinations. There are three nontechnical "rules of thumb" that can be used to determine if the work area has excessive noise.

• If it is necessary to speak in a very loud voice or shout directly into the ear of a person in order to be understood, it is likely that the exposure limit for noise is being exceeded.

• If employees say that they have head noises and ringing noises in their ears at the end of the workday, they are being exposed to too much noise.

• If employees complain that speech or music sounds seem

Figure 13-6.—Where engineering control measures are not feasible, as in this punch press operation, ear muffs can provide effective personal protection against a permanent hearing loss.

muffled after leaving work, but their hearing is fairly clear in the morning when they return to work, they are being exposed to noise levels that can cause a partial loss of hearing that can become permanent.

Control considerations. If the noise cannot be reduced by engineering means, administrative controls should be used. If this is not feasible, personal protective equipment should be used and a hearing-conservation program instituted.

Hearing-conservation programs. Administering a hearing-conservation program goes beyond the wearing of earplugs and/or earmuffs. Such programs can be complex and professional guidance is essential for establishing programs that will be responsive to the need. Valid noise exposure information, correlated with audiometric test results are needed to help both safety and medical personnel to make intelligent decisions about their hearing-conservation programs.

The effectiveness of hearing-conservation program depends upon the cooperation of employers, employees, and others concerned. Management's responsibility in this type of program includes noise measurements, initiation of noise control measures, provision of hearing protection equipment where it is required, and informing employees of the benefits to be derived from a hearing-conservation program.

It is the employee's responsibility to make proper use of the protective equipment provided by management. (See Figure 13-6.) It is also the employee's responsibility to observe any rules or regulations in the use of equipment to minimize the noise exposure.

Corrosives

Acids and alkalis are used in many industries. Chemical burns are painful and generally slow to heal. Workers cannot be too careful in avoiding accidental splashes, mists, sprays, and dusts, as well as avoiding direct contact with acids and alkalis. For most chemical burns, the immediate use of water is recommended until the chemical has been diluted and washed from the skin surface. For this reason safety showers are required in or adjacent to each workroom where such chemicals are used. See the discussion earlier in this chapter.

Rubber gloves, rubber boots, wooden soles, rubber aprons, tight-fitting safety goggles, face-shields, and acid-proof hoods are available to protect personnel working with strong liquid chemicals and should be worn when needed.

First aid. Because immediate washing of the skin or eyes with a generous amount of water is the most effective first aid treatment for chemical burns, it is essential that an ample supply of water be made available, by having emergency showers or eyewash fountains near all places where workers might be exposed to harmful chemical agents.

Different types of chemical agents produce different effects. In addition to the prime treatment of flooding the burned area with water, exposure to a particular agent may require further first aid measures. However, laymen should not attempt medical treatment which they are not authorized to give. First aid instructions with regard to chemical burns should be developed by a physician.

Toxic and flammable substances

Exposure—general. Solvents act on the skin as well as upon other substances. Prolonged or repeated skin contact with an organic solvent may lead to dermatitis. Breathing low level concentrations of organic solvent vapors may result in headache, nausea, dizziness, and impairment of the powers of judgment and coordination. Continued exposure may cause serious permanent damage to the liver, kidneys, blood, the blood-forming organs, or nervous system. Different chemicals affect different areas.

Control of hazards. Direct contact can be reduced to a minimum by use of gloves impervious to solvents or by water-soluble protective creams. Safety goggles or face shields must be used to prevent solvents from reaching the face.

Solvents should be poured without spilling or splashing and

262

should be used sparingly. Solvent containers must be kept closed and only individual cans with safety covers rather than open vessels should be used to hold the liquids (see Figure 12-4, p. 217).

Handling small quantities. Use a UL-listed or Factory Mutual approved portable safety can for handling small quantities of flammable liquids. These can range in size from 1 pint to 5 gallons, and are made in many styles with faucets, pouring spouts, or dispensing hoses.

Flammable liquid containers (usually painted red) should bear clear, legible labels that identify contents and indicate hazards. Unpainted containers of materials specifically resistant to corrosive or paint-dissolving liquids may be used. Greater safety can be attained if all flammable liquids are transferred to safety cans immediately.

Spraying operations. Since a booth is most commonly used for spraying, adequate air velocity at the face of the booth should be provided. Usually 100 to 200 feet per minute is sufficient for adequate ventilation. Smoking, welding, open flame, and use of stoves and similar heating equipment should be prohibited in paint spray areas.

Electrical equipment in spraying areas should meet the requirements of the NFPA *National Electrical Code* for such locations, and appropriate fire protection systems should be provided.

Ventilation. One of the principal methods to control health hazards, may be defined as "causing fresh air to circulate to replace foul air simultaneously removed." (See discussion on page 258.)

(a) Exhaust ventilation. The removal of air usually by mechanical means from any space.

(b) General or dilution ventilation is adding air to keep the concentration of a contaminant below hazardous levels—uses natural convection through open doors or windows, roof ventilators, and chimneys, or artificial air currents produced by fans or blowers.

(c) Local exhaust ventilation. A local exhaust system traps the air contaminant near its source so that a worker standing at the process is not exposed to harmful concentrations.

(d) Makeup air. Clean, tempered outdoor air supplied to a work space to replace air removed by exhaust ventilation or some industrial process.

Industrial Sanitation and Personal Facilities

General requirements

The requirements for sanitation and personal facilities are covered in OSHA safety and health regulations 29 CFR 1910, Subpart J—General Environmental Controls. Particular attention is invited to OSHA's regulations for carcinogens in 29 CFR 1910.1000 where special personal health and sanitary facilities are specifically required for employees working with cancer-suspect materials.

263

Housekeeping

All places of employment must be kept clean to the extent that the nature of the work allows. This was discussed in detail in Chapter 3.

Waste disposal

Receptacles used for solid or liquid waste or refuse must be so constructed that they may be thoroughly cleaned and maintained in a sanitary condition. Such a receptacle should be equipped with a solid tight-fitting cover.

Vermin control

Enclosed workplaces should be so constructed and maintained as to prevent the entrance or harborage of rodents, insects, and other vermin. An effective extermination program shall be instituted where their presence is detected.

Water supply and facilities

Potable water. Potable water must be provided in all places of employment, for drinking, washing of the person, cooking, washing of foods, washing of cooking or eating utensils, washing of food preparation premises, and personal service rooms.

Drinking fountains. Drinking fountain surfaces must be constructed of materials impervious to water and not subject to oxidation. The nozzle of the fountain must be located to prevent the return of water in the jet or bowl to the nozzle orifice. A guard over the nozzle shall prevent contact with the nozzle by the mouth or nose of persons using the drinking fountain.

Portable drinking dispensers. Portable drinking water dispensers must be designed, constructed, and serviced so that sanitary conditions are maintained, capable of being closed, and equipped with a tap.

Ice in contact with drinking water shall be made of potable water and maintained in a sanitary condition.

Open containers such as barrels, pails, or tanks for drinking water from which the water must be dipped or poured, whether or not they are fitted with a cover are prohibited, as are common drinking cups and other common utensils.

Where single-service cups are supplied, both a sanitary container for the unused cups and a receptacle for disposing of the used cups must be provided.

Nonpotable water. Outlets for nonpotable water, such as water for industrial or firefighting purposes, shall be posted or otherwise marked in a manner that will indicate clearly that the water is unsafe and is not to be used.

Construction of nonpotable water systems or systems carrying any other nonpotable substance shall be such as to prevent backflow or backsiphonage into a potable water system.

Toilet facilities

Toilet rooms must be provided in all places of employment. The number of facilities for each sex is based on the number of employees of that sex for whom the facilities are furnished. Where toilet rooms will be occupied by no more than one person at a time, there must be a door that can be locked from the inside and contain at least one water closet. Under these conditions, separate toilet rooms for each sex need not be provided.

These requirements do not apply to mobile crews or to normally unattended work locations if transportation is available to nearby toilet facilities that meet the requirements.

When persons other than employees are permitted to use toilet facilities on the premises, the number of such facilities shall be appropriately increased.

The sewage disposal method shall not endanger the health of employees.

Toilet paper with holder must be provided for every water closet. Covered receptacles should be kept in all toilet rooms used by women.

For each three required toilet facilities at least one lavatory must be located either in the toilet room or adjacent thereto.

Washing facilities

Washing facilities shall be maintained in a sanitary condition.

Lavatories. Lavatories shall be made available in all places of employment in accordance with OSHA requirements (29 CFR 1910.141(d)). In a multiple-use lavatory, 24 lineal inches of wash sink or 20 inches circular basin, when provided with water outlets for each space, is considered equivalent to one lavatory. These requirements do not apply to mo-

bile crews or to normally unattended work locations if transportation is readily available to similar nearby washing facilities.

Each lavatory requires hot and cold running water, or tepid running water. Hand soap or similar cleansing agents must be provided. Individual hand towels or sections thereof, of cloth or paper, warm air blowers or clean individual sections of continuous cloth toweling, convenient to the lavatories must be provided. Receptacles for disposal of used towels are also required.

Warm air blowers must provide air at not less than 90 F and shall have means to automatically prevent the discharge of air exceeding 140 F.

Showers. Whenever showers are required by a particular OSHA standard, the showers and facilities must provide:

- One shower for each 10 employees of each sex, or numerical fraction thereof
- Body soap or other appropriate cleansing agent conveneint to the showers
- Hot and cold water feeding a common discharge line
- Individual clean towels

Change rooms

Whenever employees are required by a particular standard to wear protective clothing because of the possibility of contamination with toxic materials, change rooms equipped with storage facilities for street clothes and separate storage facilities for the protective clothing shall be provided.

Sources of Information

American Conference of Governmental Industrial Hygienists, 1014 Broadway, Cincinnati, Ohio 45202.

American National Standards Institute, 1430 Broadway, New York, N.Y. 10018.

American Society for Testing and Materials, 1916 Race St., Philadelphia, Pa. 19103.

American Society of Mechanical Engineers, 345 East 47th St., New York, N.Y. 10017.

American Society of Safety Engineers, 850 Busse Highway, Park Ridge, Ill. 60068.

American Welding Society, 2501 N.W. 7th St., Miami, Fla. 33125.

Compressed Gas Association, 500 Fifth Ave., New York, N.Y. 10036.

Factory Mutual System, 1151 Boston-Providence Turnpike, Norwood, Mass. 02062.

National Fire Protection Association, 470 Atlantic Ave., Boston, Mass. 02210.

National Safety Council, 444 N. Michigan Ave., Chicago, Ill. 60611.

Occupational Safety and Health Review Commission, 1825 K Street, N.W., Washington, D.C. 20006.

Underwriters Laboratories Inc., 207 E. Ohio St., Chicago, Ill. 60611.

U.S. Department of Health, Education, and Welfare, National Institute for Occupational Safety and Health, Parklawn Building, 5600 Fishers Lane, Rockville, Md. 20852.

————. Division of Technical Services, Cincinnati, Ohio 45202.

U.S. Department of Labor, Occupational Safety and Health Administration, 14th Street and Constitution Ave., N.W., Washington, D.C. 20212.

————. Bureau of Labor Statistics, 441 G Street, N.W., Washington, D.C. 20212.